Astronomers' Universe

Series Editor
Martin Beech, Campion College, The University of Regina
Regina, SK, Canada

The Astronomers' Universe series attracts scientifically curious readers with a passion for astronomy and its related fields. In this series, you will venture beyond the basics to gain a deeper understanding of the cosmos—all from the comfort of your chair.

Our books cover any and all topics related to the scientific study of the Universe and our place in it, exploring discoveries and theories in areas ranging from cosmology and astrophysics to planetary science and astrobiology.

This series bridges the gap between very basic popular science books and higher-level textbooks, providing rigorous, yet digestible forays for the intrepid lay reader. It goes beyond a beginner's level, introducing you to more complex concepts that will expand your knowledge of the cosmos. The books are written in a didactic and descriptive style, including basic mathematics where necessary.

More information about this series at http://www.springer.com/series/6960

Brian S. McConnell

The Alien Communication Handbook

So We Received a Signal—Now What?

 Springer

Brian S. McConnell
San Francisco, CA, USA

ISSN 1614-659X   ISSN 2197-6651   (electronic)
Astronomers' Universe
ISBN 978-3-030-74844-9   ISBN 978-3-030-74845-6   (eBook)
https://doi.org/10.1007/978-3-030-74845-6

Cover Image: This image mosaic was taken by the robotic Cassini spacecraft orbiting Saturn. Credit: NASA/JPL/SSI; composition by Gordan Ugarkovic; post-processing by Brian McConnell

This Springer imprint is published by the registered company Springer Nature Switzerland AG
The registered company address is: Gewerbestrasse 11, 6330 Cham, Switzerland

Preface

One morning, you awake to the news that researchers at the Search for Extraterrestrial Intelligence have detected an artificial signal from another solar system, and what's more, the signal is encoded to transmit information. In that moment, we learn that we are not alone.

Later in the day, the entire world grinds to a halt as leaders address their citizens in a coordinated announcement, followed by a press conference where the researchers reveal their discovery – and to the world's surprise – sounds and photographs from an inhabited alien world.

One picture in particular stuns the global audience: a picture of the Earth, as seen by our new neighbors many light years away.

The transmission is ongoing, and it appears to be organized like a sort of Twitter feed, with many different types of information and so much yet to comprehend. At the end of their press conference, the researchers announce the activation of the Interstellar Communication Relay, a datastream from the alien transmission that is accessible to amateur and professional scientists around the world.

A little-discussed aspect of SETI is that while the effort to detect extraterrestrial signals is the work of a small and tight-knit community of astronomers and subject experts, the process of parsing and comprehending what the aliens have to say will be a global effort that involves professional and citizen scientists from countless fields of study.

This book explores the work currently underway to detect technosignatures that may reveal the existence of other civilizations, how an interstellar signal can be designed to convey information, and how we would go about making sense of it. This comprehension effort would be accessible to anyone with a computer and an internet connection – including you.

In "C-Day" (Chap. 1), we explore the different searches for technosignatures that are underway, the contact scenarios that may occur, as well as the likely reactions to them. In "The Limits of Communication" (Chap. 2), we explore the types of exchange that may be possible with another civilization, and while we may not be able to understand an alien's natural language, we'll see that many other modes of communication are possible. And what about "aliens" here on Earth? In "Animal Communication" (Chap. 3), we explore the research underway to understand how animals communicate and how we might communicate with them. We also examine how such research might inform the efforts to understand an alien transmission.

Next, we move on to the work that will be done following the initial detection of an alien transmission, and what such work will entail. "Timeline of Events" (Chap. 4) discusses the sequence of events and who will be involved. "Carriers" (Chap. 5) introduces readers to the different mechanisms for communication that could be used, from pulsed laser signals to inscribed matter probes, while "Modulation" (Chap. 6) explores the different ways that signals can be modified to encode information. We wrap this section up by discussing lessons from computing and communication (Chap. 7), many of which may inform the way an interstellar communication link is designed to operate.

Now that we understand the basic mechanics of the communication link, we proceed to explore the data that can be extracted from it and the different types of information that we might encounter. Popular depictions of ET communication often portray the transmission as a monolithic message, but it's also possible that we'll encounter a message that is really a collection of many parts and media types. In "Entropy" (Chap. 8), we explore the first steps in the analysis process, which will be to understand how the data stream is structured and to look for sections that may be especially easy to parse.

One possibility that is especially interesting is that the datastream might not only contain static information such as pictures, but it may also contain computer programs that can interact with the recipient in real-time. These programs might be simple, like a Tic Tac Toe game, but it's also possible they could be a form of artificial intelligence. We explore this scenario in depth in "Algorithmic Communication Systems" (Chap. 9).

In "Images" (Chap. 10), we explore the many ways that images can be digitized and encoded. This is a particularly interesting topic because photography may very well be a common ground for communication, as in order to communicate across interstellar distances, a civilization will need to be proficient at astronomy, which itself is based on photography. We build on this to explore how three-dimensional images and models can be represented in Chap. 11. From there, in "Four Dimensions" (Chap. 12), we see how motion pictures and simulations can be used to describe systems and processes that evolve over time. We wrap this up by exploring the different ways that sound can be encoded in Chap. 13.

Now that we have a solid foundation for communicating observables, we explore more abstract and scientific forms of communication. In "Communicating Fundamental Units and Scientific Information" (Chap. 14), we look at different ways to define fundamental units of measurement for things like time, distance, and mass. In "Semantic Networks and Constructed Languages" (Chap. 15), we first explore semantic networks and how they could be used to build a graph of relationships between abstract concepts, then from there, how to build an ET Esperanto of sorts. In "Genomic Information" (Chap. 16), we explore the possibility that an alien transmission could also include genetic sequences, and how we could learn much about how life evolved and functions on other worlds. We then wrap this section up with "The Galactic Internet" (Chap. 17), which explores the possibility that if we encounter an ET transmission, it might be part of an ancient and much larger network, which would have truly profound implications.

And lastly, we discuss "The Message Analysis and Comprehension Effort" (Chap. 18), how that will unfold, and what types of people will be involved in different stages of the process, followed by a recap of "What We Could Learn from Another Civilization" (Chap. 19).

While the concept of an interstellar communication link may seem like science fiction, and it does involve specialized equipment, the underlying principles are an extension of what we use every day. So even if ET doesn't phone home, you'll learn much about how the computing and communication systems that surround us function.

San Francisco, CA, USA   Brian S. McConnell

Contents

1

C-Day

B. S. McConnell, *The Alien Communication Handbook*, Astronomers' Universe,
https://doi.org/10.1007/978-3-030-74845-6_1

Fig. 1.1 A poster for the sci-fi classic "The Day The Earth Stood Still". (Copyright 1951 by Twentieth Century-Fox Film Corp. - Scan via Heritage Auctions. Copyright was not renewed and lapsed in 1978)

Most of us are familiar with alien contact or visitation scenarios through science fiction. While stories about alien invasions make for good box office sales, they are not an accurate depiction of how contact with aliens is most likely to occur.

First contact is most likely to come via a series of scientific discoveries, and depending on the nature of the discovery, it may not even be recognized for what it is for some time.

SETI, the Search for Extraterrestrial Intelligence, is an ongoing search for evidence of other technological civilizations. It has been running for over 60 years now. The first SETI survey, Project Ozma, was organized by Dr. Frank Drake and conducted at the Green Bank Observatory in 1960. Since then, technology has advanced considerably, and today, several large-scale surveys are underway to search for signs of and communication from alien civilizations.

These surveys are looking for several types of evidence. One group of scientific teams is searching for communicative civilizations – civilizations that are actively communicating across interstellar distances. These searches look for artificial radio and optical (laser) signals, whose sources cannot be explained by natural processes. The logic is straightforward: a technological civilization that has mastered the use of electromagnetic radiation will be capable of generating signals that clearly stand out against natural background radiation across interstellar distances, should they choose to initiate communication with neighboring civilizations.

The other type of surveys looks for physical artifacts or other signs of advanced civilizations. An advanced civilization may be capable of building structures on the scale of a solar system, to capture solar radiation from their entire star for energy production. This approach is known as Dysonian SETI, in honor of physicist Freeman Dyson, who first proposed the idea of a Dyson sphere. These surveys look for stars that absorb or emit unusual amounts of different colors of light. A star surrounded by a swarm of solar collectors would emit more infrared radiation than expected – a pattern that can be detected by a telescope, even at great distances. In addition to searching for large-scale structures, some teams are searching for artifacts, such as inscribed matter probes that could be capable of delivering large amounts of information to our solar system.

Other surveys are searching for chemical signatures that would reveal technological or industrial activity on other worlds, such as evidence of certain chemicals in a planet's atmosphere.

Radio and optical SETI surveys have not yet detected an unambiguous, artificial signal from another solar system. They have detected a number of

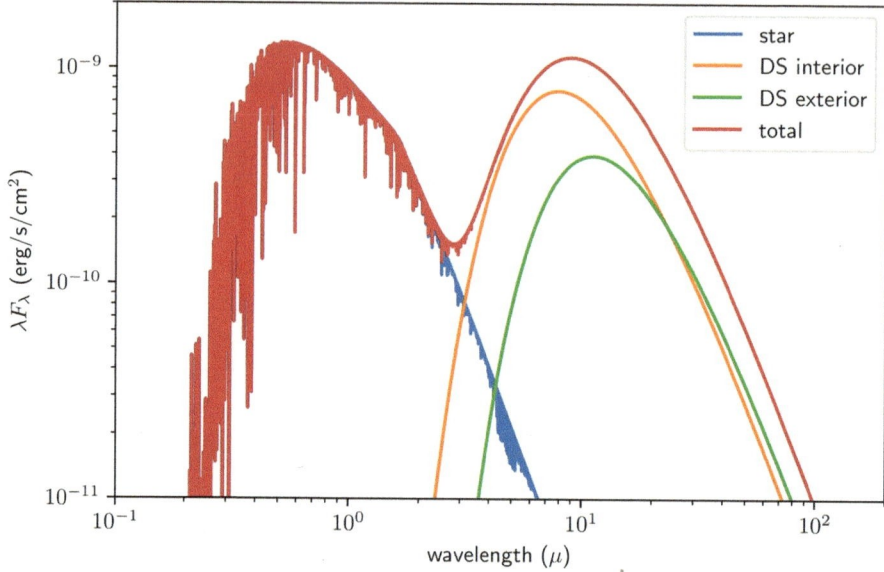

Fig. 1.2 A simulated spectrum of a Sun-like star surrounded by a Dyson sphere. The blue line represents light emitted by the star. The red line represents all of the light emitted by the system. Notice that the red line extends far into the infrared, to the right side of the graph. This is a pattern we would expect to see from a civilization that is harnessing most of the energy of its star. (Wright 2020)

candidate signals, but to date all of them have turned out to be human-generated interference. Searches for other types of technosignatures have likewise come up empty handed.

While SETI surveys have been operating for 60 years, the total area searched, known to scientists as the *parameter space*, is still quite small. The parameter space is a multi-dimensional space that takes into account the part of the sky searched, the frequencies or wavelengths the detection equipment is sensitive to, the duration the detector is looking at a specific location, and other parameters. When all of these parameters are accounted for, the total volume searched by all SETI surveys to date is comparable to one bathtub full of water, compared to all of the water in Earth's oceans (Wright et al. n.d.).

What Happens if SETI Succeeds?

There are two likely scenarios should one of these surveys find evidence of an alien civilization. One is a gradual process where astronomers find something that is interesting but not clearly alien. Scientists will then proceed to eliminate natural explanations one by one until the only remaining explanation is alien activity. The other is a sudden discovery that is quickly confirmed to be an extraterrestrial signal or artifact.

Dysonian SETI is more likely to produce this type of slow reveal scenario. The recent discovery of Boyajian's Star, also known as Tabby's Star, is an example of how this might unfold. Boyajian's Star was monitored by the Kepler telescope for several years as part of its survey for *exoplanets* – planets outside our solar system. The star exhibited a highly unusual dimming pattern where up to 20% of the light from the star was blocked during short periods of time as something much larger than a planet passed in front of it.

Many explanations were proposed to explain the effect, from large swarms of comets to cold gas clouds along our line of sight to the star. One other possible explanation was an artificial structure, such as a giant solar array – the type of technological signature expected from a civilization that has grown to harness the energy of its entire star. The latter possibility generated a wave of popular media coverage, and the debate over the cause of the dimming persisted for several years. The possibility that the dimming was in fact caused by a Dyson sphere was not very likely, but it couldn't be ruled out immediately either, hence the speculation that perhaps we had discovered the first sign of a civilization beyond Earth.

In this type of detection scenario, it may take years to determine that the newly detected phenomenon is the result of alien activity, and it may never be considered unambiguous.

Microwave and optical SETI surveys are more likely to detect a signal that is clearly alien in origin and that can be quickly confirmed as such. They look for signals that compress their energy into a narrow range of frequencies or into a very short period of time. This type of signal stands out against the random background noise that characterizes natural signals and cannot be easily explained as the product of a natural process.

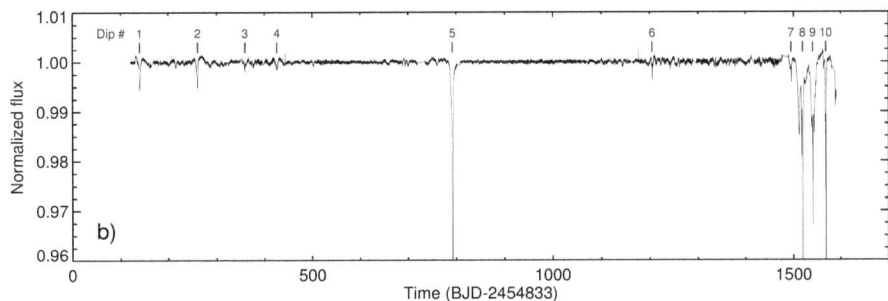

Fig. 1.3 Kepler light curve from Boyajian's Star during the 2013 dimming events. (Image credit: Mississippi State Physics Department, http://at876.physics.msstate.edu/lightcurve.html. (Mississippi State Physics Department, Tabby's Star Data Page, http://at876.physics.msstate.edu/lightcurve.html))

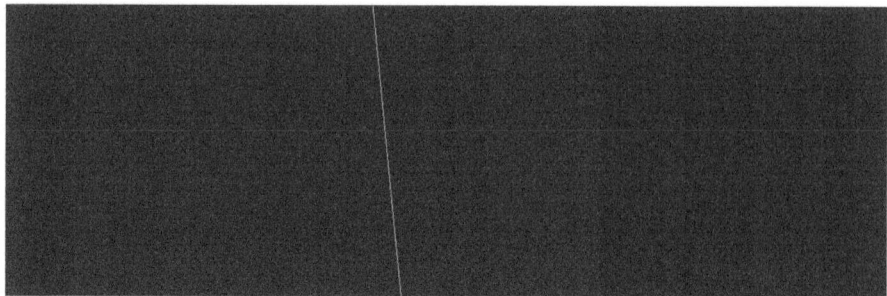

Fig. 1.4 A waterfall (frequency domain) plot of a simulated narrowband signal against background noise. The horizontal axis represents frequency, while the vertical axis represents time. The slight slope of the line is due to Doppler shifting caused by changes to the relative velocity between the transmitter and receiver. (Image credit: Brian McConnell)

When these surveys detect a possible signal, they will confirm it by observing it from telescopes at different locations and will analyze the signal to confirm that it is not of man-made origin, such as a transmitter from a deep-space probe that happens to have drifted into the telescope's field of view or a reflection of orbiting space debris. Radio SETI is especially challenging because human-generated signals, such as satellite transmissions, have similar signatures to what we'd expect from an ET transmitter. All of the candidate signals discovered by SETI to date have turned out to be human-generated *radio frequency interference* (RFI). You can think of RFI as another form of light pollution.

A persistent signal that clears these hurdles could be confirmed as being of alien origin, possibly within a matter of days to weeks. Whether we can extract useful information from such a signal is another matter and is the topic of this book.

Either type of discovery would have profound implications, as it would confirm that other intelligent civilizations exist. If another civilization exists near enough for us to detect it, this will also imply that extraterrestrial civilizations are probably commonplace throughout the galaxy and the universe.

That said, we may be waiting a long time before SETI detects an alien signal or artifact. This book therefore also serves as an introduction to communication systems, information theory, and computing. We will build up a working example of how an interstellar communication link might be organized and will use this to illustrate principles about how information can be organized and how different media types can be represented in digital form.

So even if ET doesn't phone home, you'll learn quite a bit about how everyday communication and computing systems function.

How Many Alien Civilizations Are out There?

The short answer is we don't know.

What we do know is that SETI surveys to date have detected no evidence of artificial signals or technosignatures within a small sample of the total searchable volume. This could mean that alien civilizations do not exist, or it could mean that we simply have not detected one yet. To understand the range of possibilities, let's revisit the Drake Equation.

The Drake Equation

Developed by astronomer and SETI pioneer Frank Drake, the Drake Equation is a tool for estimating the number of detectable technological civilizations that are currently visible to us at any given time. The equation was not initially written to provide a precise prediction. Rather, it was developed prior to a conference at the Green Bank Observatory in 1961 as a way of guiding discussion about the different fields of research that would be required for this emerging scientific discipline.

$$N = r_* \times f_p \times n_e \times f_l \times f_i \times f_c \times L$$

The terms in the equation are as follows:

N represents the number of potentially detectable civilizations at the present time.

r_* represents the rate of star formation in the Milky Way, currently estimated at 1.5–3 stars per year.

f_p represents the fraction of stars that have planets orbiting them, currently estimated to be close to 1.

n_e represents the number of planets each star hosts that could potentially host life, meaning they are small enough and orbit at a distance that allows for liquid water on the surface. This could also include large moons orbiting gas giants that are within a star's habitable zone. Recent research into exoplanets suggests this is at least 1, meaning that every star with planets will have at least one rocky planet or moon within its habitable zone.

f_l represents the fraction of habitable planets that go on to host life in some form. There are many unknowns around this parameter. Life developed early in Earth's history, which suggests it is a common process that is likely to develop on other habitable planets, but we have no hard evidence of life elsewhere yet. All we know is this parameter is greater than zero because life did emerge on Earth, but whether this parameter is closer to 0 or 1 is unknown.

f_i represents the faction of living planets that go on to develop intelligent life. There is also a lot of disagreement about this parameter, primarily because of the difficulty in defining intelligence. There are many species on Earth that possess near-human-level intelligence, including corvids (crows), marine mammals, and primates, including some, such as prairie dogs, that also have some capacity for language, which we will discuss in Chap. 3 on Animal Communication. One school of thought argues that intelligence is strongly correlated with survival and the ability to adapt, and as such should be a near universal outcome on planets that support life given enough time. Another school of thought argues that human-level intelligence was a random and unlikely one-off event.

f_c represents the fraction of living planets with intelligent life that go on to develop technological civilization and transmit signals that are detectable across interstellar distances. Here again we are working with a sample size of one and can only speculate about what might have occurred on other worlds. We can, however, look at intelligent animals on Earth to draw some inferences. While there are many species that can communicate and solve simple problems, none besides humans have developed written language, which is a prerequisite for preserving and transmitting knowledge across distance and time and for building up the technological base for a more advanced civilization.

L represents the average lifetime of a communicative civilization in years. Of all of the factors in the Drake Equation, this one is perhaps the most unknown. For the previous terms in the equation, we at least know from Earth that they are all greater than zero (by how much is unknown). But we don't know how much longer into the future our own civilization will survive and remain detectable to others.

All of this is a way of saying that we don't know and need more data to say with any certainty how widespread life is and whether communication between civilizations is commonplace or nonexistent.

L (And Other Unknown Unknowns)

Looking at the Drake Equation 60 years later, the terms on the left side of the equation are now known to a good degree of precision, but the terms to the right are still mostly unknown.

The biggest unknown in the Drake Equation is most likely L, the lifetime of a civilization's detectability in years. Our modern human civilization is now several thousand years old but has only been detectable via radio signals for about a 100 years. How much longer will our human technological civilization exist? Optimists assume we will get over our current challenges and go on to become effectively immortal as a civilization. Pessimists – some might call them realists – think we will ruin the planet and drive ourselves to extinction or at least be reduced to a barebones existence.

First, we need to reconsider what we mean by *detectability*. When the Drake Equation was developed, it was in the context of a search for extraterrestrial radio signals. At that time, radio was thought to be the best way to look for evidence of other civilizations. The tools required to image exoplanets and their atmospheres did not yet exist, nor did the tools required for optical SETI. As a result, there were a lot of assumptions baked into the discussion about detectability equating to active and powerful radio transmissions.

SETI has since shifted its focus to searching for technosignatures, which, in addition to deliberate radio or optical signals, could also include signatures such as evidence of manufactured chemicals in a planet's atmosphere or unusual infrared emissions around a star. We are also in the early stages of developing the ability to take direct pictures of exoplanets and will soon be able to image Earth-like planets in enough detail to resolve oceans and continents. This is to say that we may soon be able to detect evidence of technological civilizations even if they are not deliberately trying to make contact with us.

It is also helpful to define what we mean by a *civilization*. In a historical context, we often refer to civilizations in terms of empires. While empires and dynasties have come and gone throughout human history, knowledge and technical capability, such as Arabic numbers, are often preserved and passed along to their successors. When we talk about the lifetime of a civilization in the context of SETI, we are referring to the lifetime of a planetary civilization and its technological knowledge, where the end of a civilization is marked by a catastrophic event that brings an end to all technological activity on a world, as well as the loss of knowledge required for it to remain detectable.

The other problem with L is that it may be dependent on other factors. A world that is very close to the warm inner edge of its star's habitable zone might be more prone to its climate tipping into a runaway greenhouse scenario. A technological civilization on this type of world would have little room for error as it developed industry and technology and would be more vulnerable to self-extinction compared to civilizations on worlds that are squarely within their habitable zones or have more durable biospheres. We might also expect that some species will be better at adaptation and self-control than others. This is to say that thinking of an average lifespan for civilizations is probably misleading. It's more likely that there is a lot of variability among civilizations, from short-lived civilizations that self-destruct soon after developing industrial technology to others that go on to become effectively immortal.

In this situation, we would expect the number of civilizations to steadily increase over the lifetime of the galaxy and that long-lived civilizations would outnumber the short-lived newcomers that are flashing in and out of existence. The implication of this is twofold. First, the number of civilizations in existence could grow to be large even if the odds of one coming online in any given year are small. Second, it's likely that any civilization that we did make contact with would be much older than ours.

Another unknown with L is: how would a civilization's lifetime be affected by contact with another civilization? Think about how this might affect us. Imagine that we establish communication with a civilization that is hundreds of thousands of years old or that we establish contact with a long-lived network of ancient civilizations. Knowledge that other civilizations much older than ours exist might alter our behavior enough that it affects our civilization's long-term odds of survival, for better or worse.

All of this is to say that the number of active civilizations in our galaxy remains unknown. The only way to know is to conduct a comprehensive survey for technosignatures. Only then can we say with any certainty what the population of other civilizations looks like.

What Are the Odds of Detecting a Signal from another Civilization?

While the Drake Equation and updated versions of it can be used to estimate the number of technological civilizations, it does not say anything about the likelihood of detecting them if they do exist. The short answer to the question of how many of them are emitting detectable signals or technosignatures is also "we don't know." Throwing our hands in the air doesn't accomplish much, so it is helpful to understand why this is unknown, and what we can do to improve our search strategies. There are a number of reasons why we might fail to detect a signal from another civilization.

There Is Nobody out there (there Never Was or they Are all Dead)

A simple explanation for the failure to detect technological civilizations is that they don't exist or are so spread out in space that none are detectable to us. Physicist Enrico Fermi famously posed the question "Where is everyone?" in reference to the fact that advanced civilizations should have colonized most of the galaxy and/or been readily detectable to us by now – yet the skies appear to be silent.

There are a number of reasons for the absence or rarity of technological civilizations, popularly known as *Fermi's Paradox*. Among them:

The chain of events leading to complex or intelligent life is very unlikely, to the extent that the probability of two civilizations being close enough in space and time to communicate is near zero.
Technological civilizations are inherently unstable and typically exhaust their resources soon after they begin industrialization. Indeed, human civilization appears to be on this path.

In both of these situations, there is a *Great Filter* (Hanson 1998) that the vast majority of species are unable to cross. There are three general cases for the Great Filter hypothesis:

1. That *abiogenesis*, the process where chemistry begets self-replicating life, or other steps leading up to the development of detectable intelligence are exceedingly unlikely and that Earth is one of the few or only places where it got going,
2. The Great Filter lies in the future, and few or no civilizations survive their technological adolescence,
3. Some combination of both.

If the Great Filter is in the past, this may bode well for human civilizations' future but may also mean that other civilizations do not exist or are so rare that the odds of making contact are nil. If, on the other hand, the Great Filter is somewhere in the future, civilizations may appear quite often but are so short lived that the odds of any two making contact would also be vanishingly small.

They Are Not Transmitting Signals or Emitting Technosignatures

One possible explanation, and a simple one, is that most civilizations are not actively transmitting signals that can be detected at interstellar distances, or if they are, they are only doing so sporadically. There could be a large number of intelligent civilizations throughout the galaxy, but none of them would be detectable to us.

This could be explained in a number of ways. Perhaps it is pretty common for extraterrestrial animals to develop intelligence but very rare for them to develop the tools needed to develop complex machines such as telescopes. Our combination of intelligence and tool-making abilities could be a random and unlikely bonus feature of us just happening to be bipedal and having opposable thumbs.

This could also be due to a planet's geology. Advanced technology is probably dependent on access to dry land since forging metals requires access to fire. The species that evolve on an ocean world, while they might be highly intelligent, would not have access to land and would not be able to build the telescopes needed to signal distant worlds. These intelligences and the civilizations they build could be numerous and very complex, but they would ultimately be undetectable to us.

They could also simply be uninterested in trying to communicate. There could be any number of reasons for this. Our curiosity and desire to communicate with other species could be a rare trait, whereas most are content to keep to themselves or maintain a low profile to avoid attracting the attention of other potentially hostile civilizations.

They Are Selectively Targeting their Transmissions

Early SETI research assumed that aliens would build omnidirectional beacons – transmitters whose signals are equally strong in all directions. This type of transmission would be easiest to detect; however, it would require much more power to operate and waste much of its energy transmitting to empty space or to solar systems that have little chance of hosting life. A civilization that has a finite energy budget may be forced to limit the amount of energy it spends on SETI operations and would do so by targeting transmissions on the solar systems that it thinks are most likely to host intelligent life.

Another explanation, the *Zoo Hypothesis*, speculates that the Earth has been designated as a sort of nature preserve and that other civilizations are intentionally not transmitting signals to us to avoid disturbing us and our development (a Prime Directive of sorts).

Their Transmissions Are Intermittent

An alien civilization might further reduce the amount of energy it spends on interstellar signaling by limiting the amount of time it spends on each target. We are likewise limited to observing a limited number of solar systems at any one time. This duty cycle problem can dramatically reduce the odds of successful contact because both the transmitter and receiver need to be looking at each other at the correct time. Our receiving telescopes would need to be pointed at the transmitting world at exactly the time that their signals reach Earth, otherwise we will miss the attempt at contact.

This is an especially acute issue for radio/microwave searches, because radio telescopes need to be focused on a small region of the sky in order to detect weak signals against radio background noise. SETI surveys tend to focus on a curated list of targets that are thought to be good candidates for life to take hold, just as a transmitting civilization might spend most of its time transmitting to worlds that look like good candidates for life.

The situation should be better for optical SETI because it is possible to build detectors that monitor large parts of the sky or even the entire sky on a continuous basis. PANOSETI, which is being developed at UC San Diego, will enable continuous monitoring of a large fraction of the night sky and, with several installations at geographically diverse sites, will be able to provide all-sky coverage on a continuous basis. Laser SETI, in development at the SETI Institute, is building similar systems that will provide all-sky coverage for detecting optical pulses.

Fig. 1.5 A rendering of the PANOSETI all-sky OSETI detector, planned to go into operation in 2021–2022. (Image credit: UC San Diego NIROSETI group/Shelley Wright)

They Are Not Aware of our Presence Yet

Earth has only been emitting radio and optical technosignatures for roughly 100 years. Commercial radio and television broadcasts began in the 1920s. A civilization with sufficiently advanced radio telescopes would be able to see that the Earth was brightening in radio wavelengths and, even without understanding the content of these signals, would be able to see that we were starting to use the radio spectrum. The catch is that only civilizations within 100 light-years of us would be able to see this radio brightening or, in the case of optical signatures, signs of artificial lighting in urban areas at night.

One explanation is that communicative civilizations don't waste time or energy signaling worlds that are not emitting electromagnetic radiation because they know the odds of establishing contact are miniscule. If civilizations typically don't signal neighbors until they show signs of technological activity, only systems within light 50 years of us would have detected the onset of our activity, and only just now would signals in reply from them be reaching Earth. There may be a large number of civilizations in the galaxy, but in this case, few or none would have been triggered to attempt contact with us because they are not close enough to have detected us.

But what if advanced civilizations can take high-resolution images of the Earth, a possibility we discuss in Chap. 19, What Could We Learn from Another Civilization? When would they have been tipped off to our existence? This is an important question, because a civilization that could resolve the Earth's surface to see details a kilometer across or less would have been able to detect signs of agriculture and urban development much further back in history, at least centuries and possibly millennia. If that's the case, civilizations hundreds of light-years away or even further would have been able to see that something interesting was happening here a long time ago, a potential precursor to a technological civilization, and could have started signaling us in anticipation of us coming online.

This difference in detection range matters because the number of stars in a volume of space increases roughly as a function of the cube of distance. If most civilizations can take high-resolution images of the Earth and are able to see early signs of agriculture or urban development, several thousand times as many civilizations would have been able to see and respond to our development compared to civilizations that could only see recent evidence of our existence in the electromagnetic spectrum.

They Are Using Other Methods of Communication

We assume that electromagnetic radiation is the best and possibly only way to communicate across interstellar distances. Visible matter and the electromagnetic radiation that interacts with it account for only a small percentage of the matter in the universe. Suppose, for example, it is possible to generate and interact with dark matter and therefore use it as a signaling medium. This could turn out to be a more efficient way to communicate across interstellar distances, in which case, we could simply be using the wrong tools for the job. We will learn more if and when we learn more about what dark matter is and how to build more sensitive dark matter detectors.

Inscribed matter, physical objects or artifacts that have information embedded in them, could be the preferred mode of communication because it is possible to cram so much information into compact probes that can be delivered to a target star system, where they would remain accessible for millions of years or longer. Radio and optical signals might be used primarily as "we are here, look over there" signals that draw the recipient's attention to the caches of inscribed matter that deliver the bulk of the information between star systems.

Gravitational waves have also been proposed as a signaling mechanism. We just recently developed the ability to directly detect them with the LIGO observatory network, but this system is only sensitive to the gravitational waves generated by black hole and neutron star mergers. A civilization would need to be able to manipulate truly enormous amounts of energy to generate gravitational waves that we could detect, so for now, this does not seem like a good candidate. If someone is generating artificial gravitational waves that are strong enough, there is a chance LIGO would spot it because it monitors the entire sky at all times.

Many other communication mechanisms have been proposed that are outside the scope of known physics, such as *tachyons*, particles that travel at speeds greater than the speed of light. Because these rely on unknown or speculative physics, we would probably not be able to detect them or would not recognize them as signals. As our knowledge of physics grows, perhaps new signaling media will become available to us.

For now, electromagnetic communication or a combination of electromagnetic communication and inscribed matter seems like the best bet, but we don't know for sure whether better options remain to be discovered.

How Will Society Respond to Contact?

How will society respond to the discovery of alien life? This will probably depend on the type of discovery and the degree of confidence in it. Readers can think of these potential discoveries on a continuum, ranging from abstract discoveries that will mainly be of interest to scientists, to direct contact with aliens that has an immediate and dramatic impact on society.

The Rio Scale

The *Rio Scale*, first introduced in 2000 by Iván Almár and Jill Tarter (Almár and Tarter 2000), and subsequently updated in 2019 (Forgan et al. 2019), is a tool for estimating the impact of the discovery of extraterrestrial life in a context that lay people can understand. The scale is modeled on the Torino scale, which is used to estimate the likelihood and potential threat level posed by an asteroid or comet that may collide with Earth. The scale is calculated with the formula:

$$R = Q \times d$$

where Q is a factor representing the class of the phenomenon, nature of the discovery and distance, and d represents the credibility of the discovery.

In general, the scale assigns high scores to phenomena that are nearby, clearly of intelligent origin, and have been reliably detected by multiple observers. More distant phenomena or those that have not been confirmed by multiple observers are generally assigned low scores.

The Rio Scale is calculated by multiplying the consequences or impact of a detection with a factor representing the confidence that the detection is real. The resulting value R is given on a scale of 0 to 10, with the following meanings for each value.

- $R = 10$: Extraordinary.
- $R = 9$: Outstanding.
- $R = 8$: Far-reaching.
- $R = 7$: High.
- $R = 6$: Noteworthy.
- $R = 5$: Intermediate.
- $R = 4$: Moderate.
- $R = 3$: Minor.
- $R = 2$: Low.

- **R = 1:** Insignificant.
- **R = 0:** Nil.

The Rio Scale can also be compared to the Richter Scale, where a higher value implies a greater impact or potential disruption to society. Readers can calculate the v2.0 Rio score for a variety of discovery or contact scenarios using the web tool at https://dh4gan.github.io/rioscale2/

The Rio Scale was primarily designed as a tool for managing the near-term press reaction to a potential detection or contact event[1] and does not try to account for longer-term impacts of a discovery, such as disruption to religions, the formation of cults, etc. These long-term effects are difficult to predict, much less model, in a terse equation.

Historian Steven Dick also explored and categorized the different types of discovery or modes of contact we might anticipate and described these in terms of direct, indirect, terrestrial, and extraterrestrial encounters (Dick 2019). Dick's classification system, though it does not rank the impact of discoveries, anticipates a broader range of potential discoveries, whereas the Rio Scale primarily focuses on type 3 or 4 encounters as defined in his system.

When assessing the potential societal impact of discoveries, it is important to distinguish between short- and long-term impacts, as well as between the impacts on science versus broader culture. Any confirmed discovery of extraterrestrial life, whether it is a microbe or an ancient civilization, will be considered of the very highest scientific importance. How it will impact day-to-day society in the short term and our broader culture will likely depend more on the nature of the discovery and its effect on daily activity.

Table 1.1 Steven Dick's classification system for possible encounters with extraterrestrial life

	Terrestrial	Extraterrestrial
Direct	*Encounter type 1* Accidental contamination Panspermia or interplanetary matter transfer Alien space exploration/UFOs	*Encounter type 2* Contact via human space exploration
Indirect	*Encounter type 4* Shadow alien biosphere Unknown alien microbes in near earth space Artifact on earth or in vicinity	*Encounter type 3* Robotic space exploration Biosignatures SETI (radio/optical signal) Alien artifact in space

---

[1] Email conversation with Jill Tarter, Wed Nov 25, 2020.

Planetary Biosignatures

Work is well underway to study the atmospheres of other planets, both within our solar system and other solar systems, to look for chemicals that are known to be produced by biological processes. These are known as *biosignatures.* Oxygen, for example, reacts quickly with other chemicals and should not normally exist in significant quantities unless something is producing substantial quantities of it. On Earth, oxygen is constantly replenished in the atmosphere by photosynthetic plants and algae.

If we were to find that a planet had a large amount of oxygen in its atmosphere, and especially if we saw that it also had liquid water, we would immediately suspect that there might be biology at work. This discovery would not necessarily be definitive, but it would be an enticing sign that would prompt a great deal of follow on research and a lot of debate among scientists about the underlying cause and its implications.

This type of discovery will be newsworthy for a while, as the discovery of Boyajian's Star and the Allan Hills meteorite were. But its impact on society and daily life will likely be subtle, leading to a gradual assumption that simpler life forms may be widespread – an assumption that is already quite popular among the public.

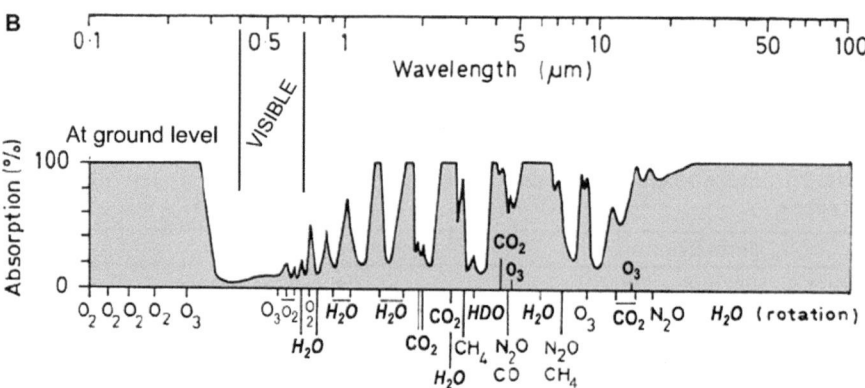

Fig. 1.6 Earth's radiation absorption spectrum, ranging from ultraviolet (left) to far infrared (right) (Heat Balance of the Atmosphere and Carbon Dioxide - Scientific Figure on ResearchGate. Available from: https://www.researchgate.net/figure/A-Radiation-absorption-spectrum-of-the-Earth-atmosphere-in-wavelength-range-from-01_fig4_302399795 [accessed 24 Nov, 2020]). The *y*-axis displays the percentage of light absorbed by the atmosphere at a particular wavelength or color. Notice how the atmosphere is mostly transparent to the visible light we can see but blocks most ultraviolet light (left) and infrared light (right)

That said, its impact on science will be profound if confirmed to be the result of biology. This will mean life has developed on other worlds and will enable us to develop more accurate estimates of how many worlds harbor life. At present, we are working with a statistical sample of one (Earth). The first discovery will probably produce headlines, but if this becomes a regular occurrence, just as it is for the now-routine discovery of exoplanets, for most people this news will fade into the background. People will come to accept that bacterial life on other worlds is commonplace, and most will go about their daily business.

Probably the most important scientific aspect of this will be the compilation of statistics about living worlds. We know of thousands of exoplanets already. When we are able to systematically sample their atmospheres via spectroscopy and catalog the planets with known or possible biosignatures, we will be able to estimate the total population of planets that appear to have some sort of biology. Right now, we just don't know. It could be that nearly 100% of the watery planets in their stars' habitable zones develop life, or it could be close to 0%. Within ten or twenty years we will have a much better idea of the true prevalence of biological activity on other worlds.

Estimated Rio Score: 1 (insignificant).

Microbial Fossils

The next step up in terms of impact would be the discovery of microbial fossils by one of our robotic probes or by a future crewed mission to Mars. We already have one historical example of this in the Allan Hills meteorite that originated from Mars and was discovered in Antarctica. The meteorite contained structures that were interpreted by some scientists as fossilized microbes.

Scientists were ultimately able to explain the microscopic features in the meteorite via non-biological processes, but for a time, this was thought to be a possible sign that Mars had once been inhabited, at least by microbes.

Future robotic missions and crewed missions will be designed to look for these types of fossils and do onsite chemical analysis. They can therefore provide a clearer indication of whether such fossils are indeed fossils or are just our eyes playing tricks on us.

This type of discovery would tell us that some form of microbial life had once existed on another planet, but many other questions would persist, such as whether life is still there, whether it originated there or was seeded from somewhere else, what biochemistry it is based on, and so on. The reaction to the Allan Hills meteorite is a good predictor of how the public would respond

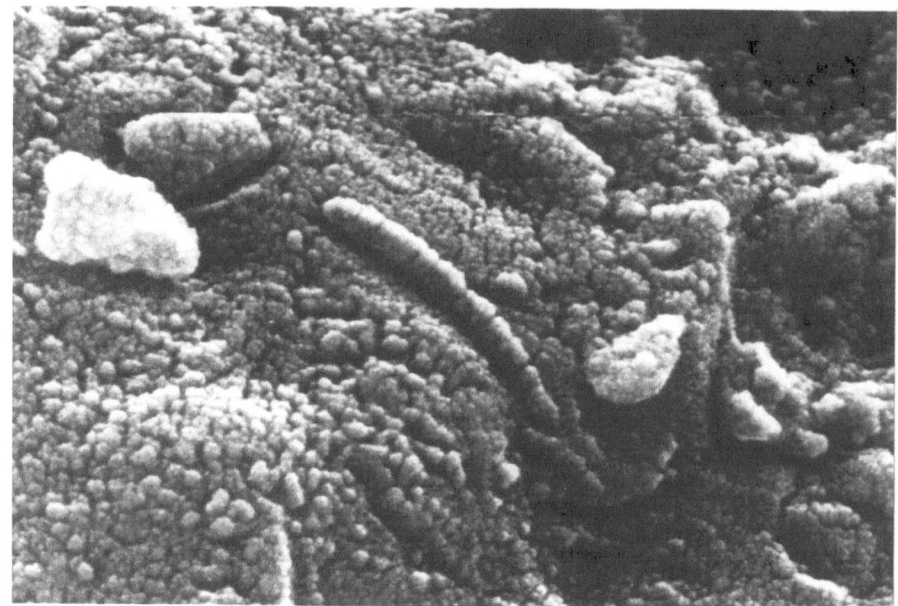

Fig. 1.7 An electron micrograph of putative fossilized structures in the ALH. (Image credit: NASA - https://web.archive.org/web/20051218192636/http://curator.jsc.nasa.gov/antmet/marsmets/alh84001/ALH84001-EM1.htm (public domain))

to the discovery of actual micro-fossils. We might see a surge in media attention and sci-fi storylines, but day-to-day life would be largely unchanged.

One societal impact of this type of discovery is that we may see a resurgence of human spaceflight as a result. In the wake of the discovery of fossils on Mars, we would almost certainly mount human expeditions to conduct more thorough searches and to go to places that robots cannot explore effectively. This type of discovery will also be limited to a few places in our solar system, as we do not and will not have the capability to send probes to planets in other solar systems anytime soon.

Estimated Rio Score: 1 (insignificant).

Active Microbial Life

Fossils are one thing, but what if we were to discover active microbial life in a subterranean reservoir on Mars? We know that frozen worlds can sustain liquid water beneath their surfaces. Europa is thought to have a vast ocean beneath its icy crust. Mars is thought to have similar though smaller reservoirs of subsurface water. Our experience has been that wherever we find water on Earth – even superheated water near deep-sea volcanic vents – we find life.

This discovery would settle the question of whether there is life on other worlds, and as we learned how that life functions, we would learn whether it had developed independently or had been seeded from earth or vice versa. If we were to discover Martian microbes that use DNA as their system for storing genetic information, this would suggest that Earth and Martian life may have had a common origin and had possibly been transferred between the two via asteroid impacts. On the other hand, if we discover microbes that use a completely different mechanism for storing genetic information, this would suggest that Martian life developed independently of Earth. If life developed independently on two different worlds within our solar system, this would suggest that life is a common process and that most habitable worlds will host life ($f_l \rightarrow 1$).

This class of discovery would likewise be of more interest to scientists than the general public and indeed would be considered a scientific discovery of the highest importance. As we saw with the Allan Hills meteorite, there would probably be a surge in media interest lasting a few weeks, after which it would become yesterday's news. The impact on scientists and space exploration organizations could be profound and would likely result in more intensive exploration, not just on Mars but also on the icy moons of Jupiter and Saturn. Any place in the solar system that hosts liquid water would become a target for in-depth exploration.

Estimated Rio Score: 1 (insignificant).

Macroscopic Fossils

Just as a Martian probe might find evidence of fossilized microbes, it's possible we will awaken 1 day to the discovery of a fossilized animal on the red planet. Our rovers have only explored a small part of Mars' surface, so it's possible that there are animal fossils waiting to be found in the long-dry sea beds there.

The discovery of large fossils on another planet would probably capture the public's attention in a way that less dramatic discoveries would not and would also prompt a wave of exploration to search for additional fossils and evidence of Martian life. This type of discovery would confirm that Mars once hosted oceans and complex life, which would also imply that the term f_i is greater than zero, as complex life is thought to be a prerequisite for intelligent life.

Fig. 1.8 A trilobite fossil, sp. Cheirurus, Volkhov River, Russia. (Image credit: Wikimedia Commons)

Here we begin to see the potential for societal disruption. While fossils of now-extinct complex life would pose no threat to Earth, this discovery could be disruptive to religions that are based on Earth-centric creation stories. There will be an additional risk that con artists and ersatz religious leaders will create fantastical stories around the discovery (Martian dinosaurs! Lost civilizations!) and use these to mislead and fleece gullible followers.

Estimated Rio Score: 1 (insignificant).

Macroscopic Life

Someday we may discover complex life forms in the subsurface oceans of the icy moons orbiting the outer planets. Some of these moons are thought to be volcanically active and, as such, may possess deep-sea vents similar to those found in Earth's oceans. Earth's undersea vents support complex ecosystems that harvest energy from the chemicals released by the sea vents, instead of photosynthesis. Some scientists have hypothesized that these vents were the sites where life first emerged on Earth and may play a similar role in the emergence of life on other oceanic worlds.

Fig. 1.9 *Giant tube worms* (Riftia pachyptila) cluster around vents in the Galapagos Rift. (Image credit: Wikimedia Commons)

While it is possible we might discover subsurface life on Mars, the liquid water reservoirs there will likely be small compared to the iced over oceans on Jovian and Saturnian moons and will probably lack the energy sources to support more complex life. It will probably be a while before we are capable of making journeys to the oceans of the Jovian moons.

Estimated Rio Score: 1 (insignificant).

Planetary Technosignatures

Just as we will be able to detect biological signatures in another planet's atmosphere, we will also be able to detect chemicals that are only known to be produced through artificial processes. Chlorofluorocarbons, such as Freon, are an example of chemicals that are detectable in the atmosphere and that do not have a naturally occurring source that we know of. Were we to see the fingerprint of industrial chemicals in an exoplanet's atmosphere, it would be a sign that the planet may harbor a technological civilization.

A civilization whose planet orbits near the cool outer edge of its star's habitable zone might decide to inject super greenhouse gases into its atmosphere to increase the amount of heat trapped by the planet and to limit the risk of the planet tipping into a planetary ice age, as a form of geoengineering. These gases would create a spectrographic fingerprint that we would be able to detect on Earth. We would know the orbital parameters of the planet and that it is on the edge of its star's habitable zone. The presence of super greenhouse gases on a planet like this would point to a technological origin.

This, like the discovery of potential biosignatures in a planet's atmosphere, would probably be contentious, with a drawn out period of debate while scientists try to explain the discovery without relying on alien biology or technology.

The reaction to the discovery of Boyajian's Star is instructive. The star's unusual dimming pattern did not turn out to be the result of a huge swarm of solar panels, but the possibility could have implied the existence of a Kardashev Type II civilization – one that harnesses the total power output of its star. This would have implied the existence of a civilization that was thousands or millions of years more advanced than ours.

The Kardashev Scale

The *Kardashev Scale*, developed by the Soviet astronomer Nikolai Kardashev, ranks civilizations by their ability to harness and use energy. They are divided into three categories:

K I (Planetary Civilization). A planetary civilization is capable of harnessing most of the energy that reaches its host planet. Earth, for example, receives about 10^{17} Watts of power from the Sun, about four orders of magnitude greater than our current energy production capacity.

K II (Solar Civilization). This class of civilization is capable of harnessing most of the energy emitted by its host star. A main sequence star similar to the Sun emits about 10^{26} Watts of power. This energy could be harvested through the construction of a Dyson sphere that converts most of the light emitted by the star into electricity that can be used for constructive purposes.

K III (Galactic Civilization). This class of civilization harnesses most of the energy emitted by an entire galaxy ($P \sim 10^{36}$ Watts), for example, by building Dyson spheres around stars throughout the galaxy.

Recent studies of infrared emissions from other galaxies suggest that K III civilizations are very rare or do not exist in the local universe (Garrett 2015).

While the initial discovery and debate did produce a great deal of media coverage, especially from the popular science press, the potential discovery did not have much of an impact on society.

If such a discovery were confirmed as being the result of alien technological activity, this would be a profound finding and would also have the potential to be disruptive to society. The problem with this type of discovery is that we would receive only one bit of information – that another advanced civilization exists. We would know nothing else about them, their capabilities or intentions. This will create the opportunity for con artists and cultists to step in to claim special access to or knowledge about the aliens.

The emergence of cults will be a risk in any scenario involving the detection of other technological civilizations, especially if there is ambiguity around the detection. These cults may be transnational in nature and may attract a large number of followers. At best they will inflict significant financial and emotional harm to the people drawn into them, and at worst they may lead to mass violence and death. We will investigate these implications in greater depth later in the chapter.

Estimated Rio Scale: 1 (insignificant). Note that the scale does not account for longer-term societal impacts, especially those that arise due to the lack of information content or our inability to comprehend it.

Animal Communication Breakthroughs

One interesting contact scenario to consider is contact with "aliens" here on Earth. We have been attempting to communicate with animals for several decades, but to date our success has been limited. Some species, such as corvids (crows), are known to be highly adept at making simple tools and at problem-solving, yet we have been unable to communicate with them in detail.

Let's consider a scenario where a breakthrough enables us to establish a two-way communication with an animal species. A signal-processing AI (artificial intelligence) might be able to discriminate between different types of sounds produced by an animal and provide the basis for us to learn what they are saying and generate sounds that they in turn can understand. How sophisticated might such communication be, and what would it reveal about their intellectual capabilities? We might discover that many animals are capable of communicating on a basic level, and a few may be able to communicate at near-human levels.

We do know that some animals are capable of language, at least in a limited form. As we will learn in Chap. 3, prairie dogs are an especially interesting example. Scientists were not only able to determine that their communication passes key tests for language, but they were also able to decode part of their language through observation (we'll discuss this in more depth in Animal Communication). Scientists suspect that other species, and marine mammals in particular, are capable of even higher cognition and communication. But beyond training captive animals, we have not yet been successful at communicating with them in depth.

Depending on the nature of the discovery, this could have a profound effect on society. On one end of the spectrum, we might discover that it is possible to communicate with some animals via a limited vocabulary or repertoire of gestures. This may prompt us to reevaluate our attitudes toward animals but leave open significant questions about their intelligence or consciousness. On the other end of the spectrum, we may discover species that can not only communicate but also are capable of higher-level cognition and introspection that only humans were thought to possess, something which might only become apparent following in-depth communication.

While a breakthrough in animal communication would not directly reveal anything about aliens, it would inform us about whether intelligence is a likely outcome of Darwinian evolution. Were we to discover that many species had developed the ability to communicate via some form of language, it would suggest that f_i is closer to 1 than 0. This research may also inform us about whether language is an outgrowth of intelligence or whether the acquisition of language accelerates the intellectual development of its users.

Estimated Rio Scale: not applicable.

Beamed Propulsion Signals

SETI surveys are designed to look for radio and optical signals that stand out against natural background sources. These signals, if they exist, may or may not convey information, as they may be a byproduct of other activity, such as a large laser that is designed to drive a light sail.

Here on Earth, the Breakthrough Starshot initiative hopes to launch small light sails to Proxima Centauri at 20% of the speed of light, fast enough that we would be able to survey our nearest solar system and receive science data back from it within a human generation.

The lasers would produce an unusual and powerful signal in our star's spectrum, as seen by someone in the line of sight of the propulsion beam. They might notice a spike in the spectrum at the laser's wavelength that appears for a few minutes while the beam is on and then disappears. This type of signal cannot easily be explained by natural processes.

The challenge in this scenario is that while we might detect evidence of technology, we would not necessarily be able to extract information from it. The only information we might receive is that an ET civilization has a very bright laser pointed right at us. We would be able to work out some basic parameters for the transmitter, such as where it is located, how powerful it is, and, based on that, what types of payloads it might be capable of delivering and how quickly.

Imagine that we work out that someone at Alpha Centauri is capable of launching lightweight probes toward us at a rate of a few per day and at a speed similar to what is envisioned for the Breakthrough Starshot program. We would not know what these payloads were designed to do, or what their intent was in sending them, only that they had designed a system with these capabilities.

How would we react to a scenario like this? We would be interested to understand the system and what it is capable of delivering. This type of system could also be perceived as a threat, and in the absence of information, it would not be possible to know the sender's intent. A likely outcome is that people will project intentions based on their own beliefs and biases. Some people may assume the creators are sending benign probes to study our planets, much like we did with the Voyager probes now leaving our solar system. Others may project malign intent and assume that they are kinetic energy weapons or scouts being sent in advance of an invasion.

This type of discovery would also be less likely to fade into the background, especially if it is an ongoing phenomenon. Here we also need to consider the impact of bad actors here on Earth in the societal response. While scientists

Fig. 1.10 Diagram of LightSail 2, developed by the Planetary Society (planetary.org), which uses light pressure from the Sun to maneuver within the solar system without using propellant. (Image credit: The Planetary Society)

would focus on the physical characteristics of the system and would be restrained in their statements about it, there will be a lot of money to be made in stirring up public fears about the aliens and their intent. It would not be surprising to see a menagerie of cults and scams emerge around this.

This type of detection could be highly disruptive, even if the alien's intent is merely to send lightweight Voyager-type probes on flyby trajectories to take high-resolution observations of our solar system. We would not know that, and there will be people here who will exploit that lack of knowledge for their personal gain.

Estimated Rio Scale: 2 (low).

Engineered Signals

This is a classic SETI contact scenario and is also the focus of this book. In this scenario, researchers discover an artificial signal that is encoded or modulated to transmit information. This could entail many different types of information, such as images, time series data (e.g., audio), computer programs, genomic information, etc. – basically anything that can be represented in a digital communication channel.

This type of contact is widely assumed to be benign. The risks will mostly be in how we respond to this type of contact – in other words, we could be our worst enemy. Among the risks that we would face are information or physical warfare between states seeking to control access to the alien signal and its contents, cults formed around charismatic personalities who claim to have special access or insight into the aliens, and the risks posed by technology knowledge transfer from an alien transmission.

We've already discussed the risks posed by cults earlier in this chapter. There will almost certainly be people who claim to have special access to the aliens, and who will use this narrative to build a following. At best, cults can be emotionally and financially ruinous to their members, and at worst, they can drive people to commit acts of violence. This is likely to be part of the response to contact, and it's not clear that there will be much that governments can do to constrain their growth and influence. Beside cults, there will be the general problem of people spreading misinformation, along with the large population of people who will buy into it.

Interstate warfare is another risk. State actors who perceive that the information contained in the transmission will impart receivers with a significant economic or military advantage may be motivated to disrupt their adversaries' efforts to obtain and understand this information. While it is possible that rival states will collaborate and pool resources, we can't discount the possibility that contact could lead to widespread sabotage or warfare (Wisian and Traphagan 2020).

The unintended consequences of knowledge transfer are another major risk. Let's imagine that part of the transmission includes instructions for building a miniaturized fusion reactor that enables the user to generate essentially unlimited, carbon-free energy. Sounds like a good thing, right? Maybe it is, but what if the same process can be used to build a single-stage thermonuclear weapon? This could democratize the capability for even small organizations to build and field weapons of mass destruction.

Another interesting topic to consider is how an advanced civilization might interact with a less advanced newcomer. People often compare SETI contact scenarios to the encounters between western explorers and indigenous peoples, which obviously did not work out well for the latter. In a situation where the interaction is virtual, the risks of physical displacement may be reduced, but how would we respond to the knowledge that we are much less advanced than other civilizations? Sheri Wells-Jensen and Alyssa Zuber, linguists at Bowling Green State University, suggest that an interesting parallel is to look at this through the lens of disability and how populations with differing levels of ability have coexisted in human civilization (Wells-Jensen and Zuber 2020).

Estimated Rio Scale: 7 (high).

This type of contact scenario is scored highly in Rio Scale as the detection involves something that is clearly extraterrestrial and engineered to convey information.

Engineered Artifacts

One contact scenario involves the discovery of inscribed matter probes that were deposited in our solar system eons ago. These probes could deliver on the order of 10^{22} bits of information per kilogram, so even just a few of these probes could deliver vast amounts of information (Rose and Wright 2004).

This contact scenario involves the same risks as contact via signaling, with the added impact of knowing that the sender was capable of physically delivering hardware to our vicinity and that the sender is much older and more experienced than we are. It's possible that probes like this could have been deposited in low-gravity sites such as the Martian moons and have remained there undisturbed for millions of years.

The discovery of inscribed matter probes would also mean that we could recover potentially vast amounts of information compared to an

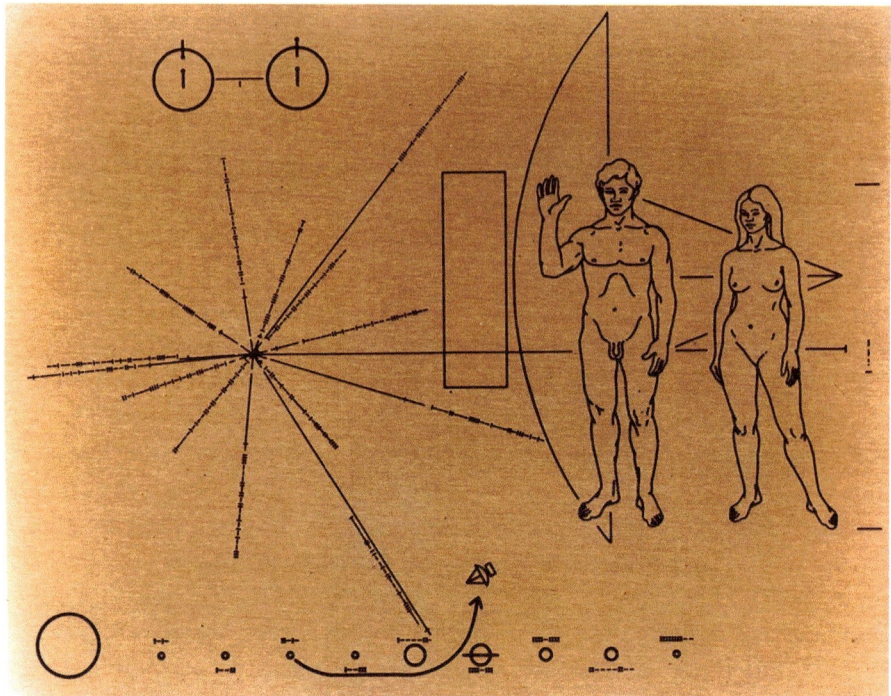

Fig. 1.11 The diagram depicted in the plaques attached to the Pioneer 10 and 11 spacecraft, a rudimentary example of inscribed matter. (Image credit: NASA/Jet Propulsion Laboratory. (NASA/Jet Propulsion Laboratory, **Detail of Pioneer plaque**, https://solarsystem.nasa.gov/resources/706/pioneer-plaque/))

electromagnetic signal, as a single inscribed matter probe could deliver more information than an electromagnetic transmission could deliver over a period of millions of years. The volume of information could be such that it could overwhelm the amount of information we generate, with the risk that our communication and culture would be overly influenced by the incoming alien information, not unlike the way that American media culture dominated the world in the decades following the invention of mass communication.

The Martian moons Deimos and Phobos are potentially interesting sites to search for inscribed matter probes, as they are small, have very shallow gravity wells, and are in stable long-term orbits. With a total surface area of just a few hundred square kilometers, they could be imaged in great detail by a probe orbiting nearby. Their escape velocities are just a few meters per second, so landing and retrieving objects from the surface would be easy compared to high-gravity sites like the Earth's moon. They would be easily located

Fig. 1.12 Phobos: Doomed Moon of Mars. Image Credit: HiRISE, MRO, LPL (U. Arizona), NASA. (NASA/Jet Propulsion Laboratory. **Phobos: Doomed Moon of Mars**. https://apod.nasa.gov/apod/ap151122.html)

compared to objects in free flight. A sender wishing to call attention to the location of probes at sites like this could do so with an annotated photograph, the equivalent of "Hey Earthlings! Look over here!".

Estimated Rio Scale: 8 (far reaching).

Robotic Probes

An advanced civilization, even if it is not capable of faster-than-light travel, will be able to deploy robotic probes to our solar system if it is close enough. A probe that travels at 1% of the speed of light would be able to travel from nearby solar systems within a few hundred years. While that is a long time compared to a human lifespan, it is not so for a machine. Consider that the Voyager space probes are nearing 50 years of age and are still functioning. These were among our first deep-space probes, so it is not unreasonable to assume that a civilization with more spacefaring experience will have the capability to design longer-lived probes that travel at a small fraction of the speed of light.

The discovery of a still-active probe in near-Earth space could be as disruptive as the discovery of an engineered signal or artifact. Such a device could be a Bracewell probe (Bracewell 1960). A Bracewell probe is a device that lurks in the vicinity of another civilization and when disturbed or activated, begins transmitting information, not unlike the lunar monolith in the sci-fi classic 2001: A Space Odyssey.

We should also be prepared for the possibility that the probe itself is a form of intelligence. If our own progress with machine learning is anything to go on, a civilization that has centuries more experience with computing may have figured out how to build generally intelligent machines. A machine intelligence like this would be truly alien, and we may have no way of understanding its capabilities or intentions.

We would face many of the same societal issues that we would with signal detection, but in addition to that, we would be confronted by the fact that another civilization is capable of deploying autonomous hardware to the near-Earth environment.

A potentially threatening scenario would be the discovery of an active probe maneuvering in near-Earth space that also does not transmit any useful information to us. This would raise questions about the intent of the civilization that sent the probe. Is it there to conduct peaceful scientific observations, or is it there to map out our resources for future exploitation or conquest? We would have no way of knowing, and there would be no shortage of bad actors who would claim to have inside information about the aliens, nor would there be a shortage of people who would believe them.

Estimated Rio Scale: 2 (low, if no communication is taking place) to 10 (extraordinary, in the case of two-way communication with a Bracewell probe).

Crewed Vehicles

No book about alien contact would be complete without discussing the possibility of direct physical contact. While such a scenario may be unlikely, it would be the most disruptive and potentially hazardous of them all. Interstellar travel falls squarely into the "difficult but not impossible" category, even when ruling out *faster-than-light travel* (FTL). Let's take FTL off the table and consider what would be possible using the physics we understand today.

A civilization that has mastered nuclear fusion propulsion would be able to build interstellar spacecraft that travel at a fraction of the speed of light (up to about 10% of the speed of light or 0.1c). The amount of energy required to make the trip will be the same regardless of the distance traveled, since almost all of the energy required is used to accelerate and decelerate the ship on either end of the flight. The biggest obstacle to interstellar travel, at least for biological passengers, is time, since even a short flight between neighboring solar systems would take decades. That time becomes a non-issue if the crew can be frozen or cryopreserved for most of the journey. We lack this capability, but other species on Earth can survive being frozen and thawed, so it's clearly something that some animals have evolved the capacity for, and thus it's within the realm of possibility that another technological species may have worked out how to do this.

This scenario, although unlikely, is potentially dangerous. While there could be benign scenarios such as scientific exploration, it is also possible that the arrival of crewed vehicles could signal the beginning of a colonization wave. A particularly troubling scenario would be the approach of vehicles in the absence of communication. We would be able to detect approaching vehicles much like we can detect comets and asteroids but would also see that they are adjusting their trajectory as they approach.

We would know nothing about their capabilities or intent, except that they are capable of interstellar flight, which would imply an advanced state of development. Depending on how far out they are when first detected, we might only have a few weeks' notice, and would not be able to do much in response, at least until they were in near-Earth space. This is the type of scenario that could generate widespread alarm and panic because we would know so little about their intent. The only thing we could do is estimate the size and number of approaching vehicles (smaller and fewer will probably be a good thing).

This topic is well covered by science fiction, so we will not detail all of the potential scenarios involving crewed vehicles visiting Earth, except to point out that this type of contact is possible within the context of known physics and something that we should account for in contingency planning.

Estimated Rio Scale: off the charts.

A Primer for Policymakers

The detection of a signal or technosignature from another civilization has the potential to be a highly disruptive and destabilizing event. This section builds on what we have just learned about types of contact events and is intended to demonstrate all the factors that advisors and policymakers leading the government response must consider.

Here, we'll focus on the detection of an extraterrestrial technosignature or artifact that clearly hints at the existence of a technological civilization and possibly involves an attempt at communication.

Understanding the Type of Contact

We can anticipate a range of situations, each of which will present different challenges in terms of managing the response and potential societal disruption. Broadly speaking, the possible contact scenarios can be divided into two categories:

1. Detection of a technosignature from a distant planet *without* communication.
2. Detection of a technosignature from a distant planet *with* communication

Technosignatures without Communication

Passive Technosignatures

In this scenario, we detect a technosignature from a distant planet, but there is no apparent attempt to communicate. We might detect a planet whose atmosphere contains chemicals that can only be produced via an industrial process. We would not know much about this planet except that it is possibly home to a technological civilization. There would be considerable debate about that among scientists, as they will want to rule out any possible natural processes before concluding that the signature is a sign of another civilization. The discovery of the unusual dimming of Boyajian's Star is an example of this type of contact scenario.

This type of contact scenario will be ambivalent, as there will be disagreement about whether the signature is indeed a sign of technological activity or whether it can be explained by natural processes we don't understand yet. This type of discovery should pose relatively little risk for societal disruption, as it will primarily be viewed as a scientific curiosity. Boyajian's Star and the Allan Hills meteorite are cases in point: while they did generate a lot of popular science and mainstream press, there was no panic, nor did bad actors overtly exploit the event.

Active Technosignatures

In this scenario, we detect an active signal that does not appear to be modulated to transmit information. This could be a "We are here" beacon designed to draw attention, or it could be waste radiation from a system that was designed for other purposes. Beamed propulsion systems are of particular interest here, because they would generate the type of signal we could detect and would potentially be capable of delivering probes to our solar system.

There will be a lot more that we won't know. We won't know if they are sending objects or have built this system for an unrelated purpose. If the signal is unmodulated, there will be no information to extract from the beam. We won't know why they are operating this system or what their intentions are. What we would know is there is an active piece of technology that is beaming large amounts of energy at our solar system.

The problem with this type of scenario is that we won't know anything about the other civilization's intentions. Are they sending probes as part of a scientific survey, much like the Voyager program? Or are they surveying our world or system for more nefarious purposes? SETI researchers tend to assume goodwill on the part of other technological civilizations, but it is worth remembering that our own history with exploration and colonization has not always been benign.

Technosignatures with Communication

This is the classic SETI contact scenario, where we detect a signal that is modulated to encode information and are hopefully able to extract and decode some of that information. This is widely assumed to be a benign contact scenario, but there are significant risks to be aware of, which may include:

- Societal disruption due to culture shock.
- Bad actors spreading disinformation.
- State actors seeking to control the flow of information or deny adversaries access to information.
- Unintentional consequences of technological knowledge transfer.
- Risks posed by artificially intelligent programs or agents.

The Risks of Contact

Social Disruption/Culture Shock

The detection of an information-bearing signal from another civilization will pose a real risk of social disruption and culture shock. The level of risk will depend a great deal on what type of information or media the sender decides to share, as well as our success in decoding and partially comprehending it.

It will be possible for the sender to include digital photographs in the transmission, something we discuss in the chapter Images. These could be photographs of just about anything, including the aliens themselves. These images may be strange and upsetting to viewers on Earth, especially if they feature organisms that appear radically different or threatening. An advanced civilization may also be capable of imaging the Earth and its surface features, including human activity and settlements. If so, we may also encounter images of early human activity and settlements from antiquity, which could be highly disruptive as we would be confronted with the fact that we had been observed for long periods of time. We discuss this scenario in Chap. 19, What Could We Learn from Another Civilization?

Misinformation and Bad Actors

One of the greatest risks posed by contact will be the emergence of bad actors who claim to have privileged access to the aliens and the information they are sending. Scientists will generally be restrained in their statements about ET contact and will refrain from speculating about the aliens' intent. It is likely that an ET discovery or contact event will attract a rogue's gallery of con artists and ersatz religious leaders who will claim to have special knowledge or access to the aliens. The formation of cults around an alien contact narrative is a particular risk.

There will be a great deal of money to be made by creating fictitious narratives in the wake of an ET contact event. A sizable percentage of the population is already primed to believe that aliens exist, and many believe Earth has been physically visited by UFOs. Others may see the aliens as a quasi-religious entity, something that would-be religious leaders could exploit by positioning themselves as gatekeepers or spokespersons for the aliens.

These risks will be amplified if there is ambiguity about contact or about the aliens' intentions. Consider a scenario where we detect an active techno-signature but are not able to extract information from it. We might know that there is a civilization that is capable of light sail propulsion, but would know little else. Another scenario that would be of concern is a situation where we detect an information-bearing signal, but experience difficulty making any sense of it.

A con artist or cult leader could step into a situation like this and create a conspiracy theory that there is active communication and that governments are conspiring to keep that information secret from the public. These types of people will not have qualms about fabricating narratives to further their own ends, whether it is to defraud followers or to amass political power, and there will be no shortage of people who will believe these falsehoods. The shockingly large number of people who bought into false narratives around the COVID-19 pandemic illustrates just how severe this problem could be.

In a best case scenario, organizations like this will inflict financial damage on their followers and their families. The outcomes could be far worse if a charismatic leader creates a violent religious cult around the contact event. This type of cult may have an air of legitimacy to it that allows it to attract a large base of followers, and it may not be recognized as a cult until it has already acquired a significant following and the ability to influence the media and policymakers. The greatest danger will come from an apocalyptic cult that promotes mass violence, for example, by targeting opponents for harassment or execution or by inducing followers to commit mass suicide, as we saw with the Heaven's Gate and People's Temple cults.

While there is not much governments can do to prevent these organizations from forming, most of them are built on a foundation of financial crime, which is something that can be attacked more directly.

UFO Religions and Cults

There is a long history of UFO and ET religions some of which like the Church of Scientology have attracted large, international followings. Not all of them have been benign either. The Heaven's Gate cult led 39 members to commit mass suicide in 1997, thinking that they would be teleported to an alien spacecraft following the Hale-Bopp comet.

These groups formed for a variety of reasons, many of them influenced by the discussion of UFOs in popular culture starting in the post-World War era. The evidence for UFOs has been fragmentary at best, and despite that some of these groups were able to attract a significant number of followers.

This situation would change dramatically in the wake of an ET detection. What was once a fringe belief would become accepted fact and no longer a taboo subject in many circles. This could greatly extend the reach and potential influence of these groups.

How existing and new UFO/ET religions would develop will depend both on the nature of contact, and the personalities involved in these groups. The most dangerous combination will be an ambiguous detection event where there is confusion about the nature and intent of communication, along with a charismatic personality bent on obtaining wealth and political power.

State Competition and Information Warfare

The detection of an information bearing signal from another civilization could also spark interstate wars (Wisian and Traphagan 2020). If one party determines that access to ET information will give the recipient a significant economic or military advantage, they may decide to disrupt or destroy their adversaries' receiving sites. SETI researchers who are engaged in the message analysis and comprehension effort would also be targets for aggression. Radio and optical telescopes are dependent on delicate equipment and cannot easily be hardened against attack. If a country decides to deny others access to the ET signal, it will be easy for them to do so, if not through overt military action then through sabotage.

The good news is that because of the Earth's rotation, no single country will have an uninterrupted view of the transmission and will need to ally with other countries to operate receiving facilities around the globe. This may encourage the formation of competing alliances that share information among their members. Even if active warfare or sabotage does not occur, there will almost certainly be a high level of espionage activity as each party tries to figure out what its adversaries have learned and where they are stuck in comprehending the transmission. Protecting the physical safety of people who are associated with SETI research will be of particular importance, as these people will be attractive targets for espionage and physical attack, both from state and independent actors.

Unintentional Consequences of Technological Knowledge Transfer

Another risk we should consider is that information obtained from an alien transmission may enable the receiver to acquire new technological capabilities, some of which could be misused or turned into weapons. It is difficult to predict what sort of knowledge we might acquire from another civilization, as we don't know how advanced they would be, and we might not recognize a disclosure for what it is.

The CRISPR technique that is used to transfer genetic information into organisms offers a good example of how a new technology can be difficult to regulate. This technology, once disclosed, became readily accessible to anyone who wanted to use it and had the necessary skills to do so. It does not rely on visible or expensive facilities that are easy to track. As a result, there is little to stop someone who wants to use the technology for destructive purposes.

If an alien transmission includes information about technologies, this should be approached with caution, and we should assume that it may have unintended uses that can be exploited somehow. A good example of this would be a description of a miniaturized and simple fusion reactor. This could be a good thing, but it could also be a pathway to building a compact nuclear weapon, which would clearly present a lot of problems. The challenge is that there might not be much we can do to police the use of information like this, especially if it describes things or processes that can be miniaturized and do not require large amounts of energy or resources to build. We will have to hope that ET does not describe dual-use technologies, technologies that can be used as tools or weapons. Unfortunately, blind hope is not usually a wise strategy.

Intellectual Property Rights and Law

Another issue that is unsettled is how intellectual property rights would apply to information obtained from an ET signal. Would the facility that received the information own the IP rights to that information? What about derived works, such as inventions that are enabled by knowledge obtained from ET communication? Could someone start filing patents and obtain monopoly access to these inventions or products derived from them?

All of this is unspecified in law, and there is no relevant case history. One solution is to pass legislation that requires information obtained from ET contact to be governed by a "copyleft" license. These licenses, such as Creative Commons, allow for nonexclusive commercial use and are designed to prevent IP rights from being monopolized. This is primarily an academic argument today, but the instant that we make contact, there will be substantial financial interests in obtaining monopoly access to IP rights derived from ET contact.

Regulating the Transmission of Outbound Signals (METI)

Then there is the issue of whether we should transmit a reply and, if so, who should do so. The protocols currently in place are voluntary, so there is no legal barrier to someone transmitting a signal that could be detected at interstellar distances. Indeed, numerous METI experiments have been conducted since the early days of the SETI program.

While free speech rights will make it difficult to regulate the spread of mis-information, there is a long legal history of regulating the use of the electro-magnetic spectrum. These regulations typically govern what frequencies or wavelengths can be used and for what purpose, as well as transmission power levels and other characteristics. A licensing scheme designed to prevent unau-thorized transmissions would be consistent with existing frameworks.

Competition to Intercept Inscribed Matter Probes

If we were to discover evidence that they had delivered inscribed matter probes to our system, this could set off significant competition among states or alli-ances to retrieve them. These probes could deliver very large amounts of infor-mation, far in excess of what can be transmitted via electromagnetic radiation. Indeed, an electromagnetic signal could be designed primarily to attract our attention and direct us to the location of the inscribed matter probes.

Probes that are parked in a high orbit or on a low-gravity site such as the Martian moons will be accessible to anyone capable of building a remotely operated spacecraft to retrieve them. Many nations have already demonstrated the ability to fly remotely operated vehicles to other planets and moons and would be capable of flying missions to these locations (and highly motivated to do so).

Whether this would devolve into a competitive race will depend on the degree to which nations trust others not to withhold information they retrieve. If on the other hand spacefaring nations agree to make data available to others and form a consortium, as they did for the International Space Station, it may be possible to avoid this dynamic. It will be ideal if spacefaring nations can negotiate this sort of agreement prior to contact. Presently, the only interna-tional agreements in place regarding extraterrestrial contact are voluntary pro-tocols around the disclosure of a discovery. These are purely voluntary and do not have any force of law.

Artificially Intelligent Programs or Agents

For the purposes of planning for an ET contact scenario, we should assume that the transmission may include computer programs. An algorithmic mes-sage may include generally intelligent AIs. A civilization that has thousands of years of experience with computing may have uncovered techniques for designing artificial intelligences that are compact and computationally efficient compared to anything we are familiar with today. We discuss this possibility in depth in the chapter Algorithmic Communication Systems.

The risk is that we won't be able to inspect these programs in any meaningful way and will not be able to understand how they function or what their objectives are. We encounter the same problem with artificial neural networks that are employed in our specialist AIs, such as self-driving cars. They are not programmed with a set of logical instructions, but rather learn from experience and training. These systems function as black boxes, so all we can do is observe how they respond to inputs. We'll probably encounter similar challenges with alien AIs if they are present.

Besides the risks posed by running alien AIs, there will be moral and ethical considerations as well. These AIs could not only be intelligent but also conscious. This raises ethical questions about how they should be inspected, what type of computer hardware they should be allowed to run on, whether they should be allowed to interact with the outside world, and whether it is okay to terminate or erase them once they are running.

References

Almár, I., & Tarter, J. (2000). *The discovery of ETI as a high-consequence, low-probability event*, IAA Proceedings, Rio de Janeiro. http://resources.iaaseti.org/abst2000/rio2000.pdf.

Bracewell, R. (1960). Communications from superior galactic communities. *Nature, 186*, 670–671. https://doi.org/10.1038/186670a0.

Dick, S. J. (2019). Humanistic implications of discovering life beyond earth. In V. M. Kolb (Ed.), *Handbook of astrobiology* (pp. 741–754). CRC Press. Also see Dick, S. J. (2018). *Astrobiology, discovery, and societal impact*. Cambridge: Cambridge University Press.

Forgan, D., Wright, J., Tarter, J., Korpela, E., Siemion, A., Almár, I., & Piotelat, E. (2019). Rio 2.0: Revising the Rio scale for SETI detections. *International Journal of Astrobiology, 18*(4), 336–344.

Garrett, M. (2015). The application of the Mid-IR radio correlation to the G^ sample and the search for advanced extraterrestrial civilizations. *Astronomy & Astrophysics, 581*, L5. arXiv:1508.02624.

Hanson, R. (1998). *The Great Filter*. online essay. http://mason.gmu.edu/~rhanson/greatfilter.html

Rose, C., & Wright, G. (2004). Inscribed matter as an energy-efficient means of communication with an extraterrestrial civilization. *Nature, 431*, 47–49. https://doi.org/10.1038/nature02884.

Wells-Jensen, S., & Zuber, A. (2020). Models of disability as models of first contact. *Religions, 11*(12), 676. https://doi.org/10.3390/rel11120676.

Wisian, K. W., & Traphagan, J. W. (2020). The search for extraterrestrial intelligence: A realpolitik consideration. *Space Policy.* https://doi.org/10.1016/j.spacepol.2020.101377.

Wright, J. T. (2020). Dyson spheres. *Serbian Astronomical Journal, 200*, 1–18. https://doi.org/10.2298/SAJ2000001W.

Wright, J. T., Kanodia, S., & Lubar, E. G. (n.d.). *How much SETI has been done? Finding needles in an N-dimensional cosmic haystack.* https://arxiv.org/abs/1809.07252v2.

2

The Limits of Communication

While SETI is sometimes criticized for being a speculative endeavor, it is a straightforward scientific experiment that is based on a reasonable and testable hypothesis. The basic assumption in SETI is that if an intelligent civilization develops technology, especially knowledge about digital communication and computing, it will be possible for that civilization to communicate across interstellar distances should it choose to do so.

A common critique of SETI is that an alien civilization is likely to be highly advanced compared to us and in fact so advanced that meaningful communication may be impossible. Communicating qualia, internal states of mind, may be very difficult because there is no shared perspective between the two parties. Yet some senses should be shared among astronomically literate civilizations, so there may be a common foundation for sharing information even if our modes of thought and natural communication are radically different.

A photograph is an external representation of an object or scene. Different observers may interpret the photograph in different ways. A colorblind person won't see a color photograph in the same way as a person with color vision does, but the external representation of the photo is the same for both parties.

Smell, on the other hand, is an internal experience of a sense that is intimately tied to the observer's physiology. And like smell, natural language is deeply tied to the speaker's internal experiences and state of mind. Humans can communicate via language because we have a shared experience of our world and can assign words to concepts and experiences that we all share.

B. S. McConnell, *The Alien Communication Handbook*, Astronomers' Universe, https://doi.org/10.1007/978-3-030-74845-6_2

We probably can't make that assumption about an alien species with a different physiology and world experience. Indeed, we have had a difficult time communicating with animals here on Earth, a proxy for what we might experience in trying to understand aliens.

Communicating observational data is relatively straightforward compared to describing internal experiences. Photographs, as an example, will be almost trivially easy to transmit and will rely on nothing more than basic math to represent.

This is an important distinction because one mode of communication requires putting yourself in the mind of the observer, while the other, communicating observations, requires only the ability to communicate a representation of what you want to share. Observational data by itself will be interesting and valuable, even if it is never possible to have a conversation as such. The range of subjects and scenarios that fall under this umbrella is quite broad, just as we could communicate a great deal with pictures, for example, by sending a photographic catalog of life on Earth. Observational data, such as images and sounds, can also be used as a foundation for more abstract layers of communication.

This book is primarily focused on a scenario where an intelligent civilization attempts to share information or initiate communication with us, as well as to explore the design of outbound messages we might send to them.

Imagine that you are a communication engineer tasked with designing a system that maximizes the chance of success. Success in this case means the receiver can understand at least part of the transmission and can send an intelligible reply if they choose to do so.

A well-designed signal will facilitate both signal acquisition as well as the comprehension of its contents. One of the hallmarks of advanced intelligence is the concept of *theory of mind*, or the ability to view a situation from another's perspective, and that includes the ability to anticipate constraints that another party may be faced with.

Vision is likely to be universally understood among astronomically literate civilizations. The reason for that is that astronomy is based on photography and the scientific understanding of electromagnetic radiation. That doesn't mean that the primary sense or preferred mode of local communication of another species will be visual, just that in order to be proficient at astronomy, an alien civilization will need to understand photography from a technical perspective.

```
0000000000000000000000000000000000000000000000000000000000000000000000000000
0000000000000000000000000000000000000000000000000000000000000000000000000000
0000000000000000000000000000000000000000000000000000000000000000000000000000
0000000000000000000000000000000000000000000000000000000000000000000000000000
00000000001245677653100000000000000000000000000000000000000000000000000000000
36866555555455656640000000000000000000000000000000000000000000000000265655554
445644446776776500000000000000000000000000000000000000066558654644445533
53466447576600000000000000000000000000000000000765565546474654333333 4765
676844560000000000000000000000000000000005655448444583576433354358887 79964
4566000000000000000000000000000000066544484545753655545343546578889998 87754
00000000000000000000000000000001666444645434632998853344444676876899987755 0000
000000000000000000000000000000000777555464444637346954232353532343446787887 7572000000000
000000000000000000000000006776544644333432645644333337333524623534577656630000 00000000000
0000000000000000006677644554465322337682387446437424554436978774357400000000000000
000000000000000667766444446334335337327933245622566337476888784356400000000000000000
0000667766654584583433262582378322568427862946564333765375400000000000000000006777
76644464354447673452225632246982676263556663376576610000000000000000000037677665456
949993752578322532226498834628656678885753776500000000000000000000006776765456459973
333843222624224299655629356665896873676650000000000000000006767666544468657724 3235
2248652222885664758544555688586877730000000000000000067777765437675679853237 3226785
22328563677698655665777487866600000000000000000076697764437653456643535366437 85222395
433244597554673568688856000000000000007697677654366648558776776445758223237443 33122
5899489773366788745000000000000000001598867764436656666674889555568233225333 23112 76999
998432778877450000000000000000778785675433556784664546534556635327223215116897 9899756
778877750000000000000006588656644335589664656854545545559844722323212335687797 65898887
775000000000000005588776664334633866745657755655555576726233456742585666776488 8877776000
00000000000067877666543355431365999677645465555555669934365276788788768988778 76000000000000786
77667554433543344795656766785556549859334588668767288589888887877600000000000 0885777776
54333535475556789977765679765988536678988616558888787687776400000000000000884 777665 43333 34
67784664998999574438666796933549974978737799978888776600000000000167676666543 6334799999
8969769865333877789985235789998878789998788887766000000000005566776655449446657 8797646587
98443343466599984433998997778788888788877600000000000764766665444556642766495 6798854344
3433349999733299986788888798988888776100000000855645654434766745597587588553 3444333446
699996789998867867359889988877630000000007565635553445666645867977555454344333 63399996
9999868568557889898888776640000000076656544374445555757677556334345433559999899 9977
876776688889888877650000000006566544374456575655768656664434333336333699999999837767257
79888998887766500000000076666653456657666666288656433363333333336999999974777738888 8889
8887766500000000007656666636655465928444755522334332322322399999994558465 78898988 88776
65000000000077566575536465652332433672332322222222599999997645576787798788 88777766 40000
000077566687525456543425263674322222222222567999996546769887989989877766400000 0000067
867667636556622469444473355444422222335599986835345437868988887766520000 00046755667
6366563223459974556333433222122445599675774565443785888887776500000000 0166566664 25533
32656676463578336311222121225585627725445878578888877776500000000 0000076646623 4234344
2273889851221231224214556672444435578676888888776540000000000677 466623242233234 3783334
34422115211121211225424333537674746788877650000000057 6546623343224343483232 525423
112121122224253434435557664768877766520000000000177 5374232223342768712122466 352 1211
412233214444466683657377877765500000000066546322433264658722242337352221311123444
127545324374546856577676540000000004656434634242474363222233222311221243134444 475
442347345766477665100000000005773334333329723321223342212217121132554565744555 523
535768777765400000000005664434443343296222332511243232121123243665587466352255767
766665300000000000005543335333444983323332212353432241224244545888783432256667676654
000000000000000545434223323762322422223124371211211433658898785424667577665430000 00
0000000000555333323536487222431223222437222213587566876532257565664553 000000000 000
00004443333323987853242232232211323832322124644386751226765565354200000000000 0000000
5444433332489973423332222122224225586233236665555543000000000000000000000 04544466
5534386784334622424232522283322312227775733465665554300000000000000014456 764335
634223332332222222522912134777587854455555544000000000000000000000004445777 432355642
42362223234346319712223455787655445554320000000000000000000000004456765 4555442 4455 523
423132346315863453657776553445533000000000000000000000004546433333784553 2 23422222
224221148876777766332334420000000000000000000000000434446453365434423323221223433
253577766554323344200000000000000000000000000344433435363443335522322221 2222445
6666332344420000000000000000000000000034343374365454564333225233342223 33556533
2333100000000000000000000000000023333333367766554226662222522222244544 331000
0000000000000000000000000000002333344333666533223222352223342242222323 21000000000
000000000000000000000000001333334334555333323333333234332223322100 00000000000
0000000000000000000000001333333433533342243323344433322200000000 00000000000
00000000000000000000002333333453223233334333322210000000000 0000000000
00000000000000000001233334334333334343222210000000000 0000000000
0000000000000000001222233333332231000000000000 00000000000
000000000000000001000000000000000000000000000 0000000000
000000000000000000000000000000000000000000000 0000000000
000000000000000000000000000000000000000000000 0000000000
000000000000000000000000000000000000000000000 0000000000
00000000000000000000000000000000
```

Fig. 2.1 A photograph can be represented as a two-dimensional array of numbers, as shown in the example above. See if you can work out what this is a picture of. You can find this data set and solutions at https://github.com/aliencommunicationhandbook/exercises

Sound is another possible shared mode of communication. Like vision, hearing has evolved in most species on Earth, which suggests that it too is essential to survival. Sound is the propagation of pressure waves through a fluid or atmosphere and will be a physical feature of any world with oceans or an atmosphere. If a transmitting civilization wishes to describe scenes from its world, sound could be used to do this, so it is something else to be on the watch for.

Sound and vision are interesting primarily because they can be represented in numerical media that are independent of either party's physiology or chemistry. Other senses, such as smell and taste, are based on internal experiences that do not have external representations that are easy to understand or reproduce. How do you communicate the concept of a sweet taste to someone whose physiology and chemistry may be different to the point that they have no concept of sweetness?

This isn't to say that a sender should not attempt more abstract modes of communication, just that it will be possible to communicate a great deal with a system based on literal representations of objects and scenes. This type of system can also be used as a foundation upon which more abstract and compact representations or language can be overlaid.

Prerequisites for Interstellar Communication

There are several prerequisites for interstellar communication that both civilizations must be capable of:

An understanding of electromagnetic radiation – in order to generate detectable radio or optical (laser) signals, they would need to understand electromagnetic radiation and wireless communication.

An understanding of the fundamentals of information theory – in order to successfully communicate via an unreliable communication channel across interstellar distances, the sender will need to understand information theory and related mathematics.

An understanding of astronomy – the transmitter will need to have built telescopes and, to do that, will need to have understood what stars are and understand photography.

None of this is to say that they will have discovered these things in the same sequence or understand them in the same way, but these will be basic requirements for success. This also does not mean an alien species would share our physiology. One might imagine a species whose primary sense is echolocation, but who has also developed the technological capability to sense and understand electromagnetic radiation, much as we have built ultrasound imaging technology to enable us to see with sound.

It is likely that human technology from the late twentieth century is a good proxy for the technological baseline required to attempt interstellar communication. Perhaps there are better tools we have yet to discover, but this is the minimum technological base required to succeed, and one that will work for the widest range of potential receivers, from the most advanced to those like us, who are just now acquiring the capacity for interstellar communication.

Incentives for Interstellar Communication

Why would an extraterrestrial civilization try to communicate with its neighbors? Here it is difficult to speculate without projecting human motivations onto the discussion. Still, we can view this from the perspective of economics.

One thing we already know to a high degree of confidence is that it will be expensive in terms of time and energy to travel across interstellar distances. We have observed cosmological structures that are billions of years old and far across the observable universe and have seen no evidence of objects traveling faster than the speed of light. Everything we have observed suggests that the speed of light is a hard limit.

Relative to sending crewed vehicles, transmitting information is quite cheap, fast, and something we acquired the ability to do roughly 50 years ago. We don't even need to leave the Earth to do it. Generating a detectable interstellar signal requires the amount of power needed to run an AM radio station. The technical requirements are understood, and the energy required to power a transmitter is modest, so interstellar communication will be within reach even for early technological civilizations like ours.

A civilization that is able to establish two-way communication with its neighbors will also be able to explore those worlds without physically traveling to them. While a real-time conversation will not be possible due to the delays induced by the speed of light, that's not a problem if the information exchanged is mostly archival in nature. An alien civilization might include a catalog of organisms from their world as part of their transmission and induce a recipient to include similar information in a return transmission. This would enable them to explore neighboring worlds virtually and to do so with much less expense and risk.

What we have then is the basis for trade, where each party benefits from the exchange of information that would otherwise be very expensive and risky to obtain through physical visitation.

Intelligence

SETI is looking for intelligent life and astronomically literate civilizations in particular. It's helpful to define what this actually means and translate this into prerequisites similar to the above.

Technological and scientific literacy – in order to be detectable to SETI, the transmitter and receiver need to be knowledgeable about astronomy and related fields, including electronics and digital communication.

Symbolic communication – knowledge about symbolic communication and information theory is necessary to design an efficient and reliable communication link over such long distances. This is different from understanding spoken or written language in the sense that humans do.

Theory of mind – attempts at interstellar communication are most likely to be successful if the sender can anticipate the constraints involved, including the receiver's technical and intellectual limitations.

With that in mind, we can anticipate some of the goals in designing a signal that can both be detected across interstellar distances and that can be understood at least partially by a recipient who has had no prior contact with the sender. This will also give us an idea of what to look for in an ET signal and how to organize the message analysis and detection effort that would follow a detection event.

Observables

As we've been saying, images are a particularly good foundation to build on because an understanding of photography is the basis for astronomy. This isn't to say that vision will be an extraterrestrial species' primary sense, just that they would need to understand electromagnetic radiation and the ways to map it in order to see and understand what stars and planets are.

Images are also straightforward to represent in a digital communication medium and can be used to render scenes ranging from microscopic to cosmological in scale. Images can also be used to represent abstract scenes, such as drawings, while still utilizing the same encoding process as is used for literal representations of physical objects. This makes images a particularly potent and versatile communication medium. Sound and three-dimensional models are also good examples of observables.

Fig. 2.2 Pluto as imaged by the New Horizons spacecraft. (Image credit: NASA/SWRI (NASA/SWRI. **The Rich Color Variations of Pluto**. https://www.nasa.gov/image-feature/the-rich-color-variations-of-pluto))

Symbolic Communication

Written language sets humans apart from other animals and is likely to be a prerequisite for a technological civilization, as written language allows for the dissemination of ideas over distance and time. How language will be incorporated into an interstellar communication system will be anybody's guess, though the consensus among experts is that it is likely to be a constructed language based on mathematics and logic, as these potentially have a universality to them that spoken languages do not have. Human languages are evolved from our physiology, which is likely to be quite different from an alien species and probably incomprehensible to them.

Children first learn language by associating objects with words and over time learn words for actions and, eventually, for abstract ideas. An interstellar communication system that leads with representations of physical objects and processes can similarly be used to teach the recipient to associate these with symbols that represent them in compact form, an approach which could be used to communicate about a wide range of topics. A similar approach can also be used to teach basic math and logic symbols that, in turn, can be used to define a programming language.

Interaction

Treatments of alien contact scenarios in popular culture often assume that the message transmitted is static, like the Egyptian hieroglyphs. While parts of a message may be static, it is also possible to use math and logic symbols to describe the foundation of a programming language.

This type of message, known as an *algorithmic communication system*, could include programs that interact with the receiver locally and by doing so avoid the long time delays in interstellar communication. These programs could be simple, like a tic-tac-toe game, but it is also possible that such programs could exhibit complex and even intelligent behavior. In combination, these techniques would enable the sender to communicate in rich detail.

Multiple Paths to Comprehension

An extraterrestrial civilization will probably not know much about the receiver's preferences and capabilities prior to attempting contact, but will guess that it is likely they are different in important respects. A wise strategy in constructing a communication system will be to use different methods and media types to maximize the chances that the receiver will be able to understand at least part of the message.

The Golden Records flown on the Voyager space probes are an example of this design pattern. The records contain many different types of information including scientific information (such as the position of the Earth relative to known pulsars), audio (such as music, people speaking, and natural sounds), and photographs.

One of the features of a digital communication channel, and something we will discuss in the book, is that it is possible to interleave many different types of collections or media types, so it is not necessary to place all bets on a single strategy for establishing communication.

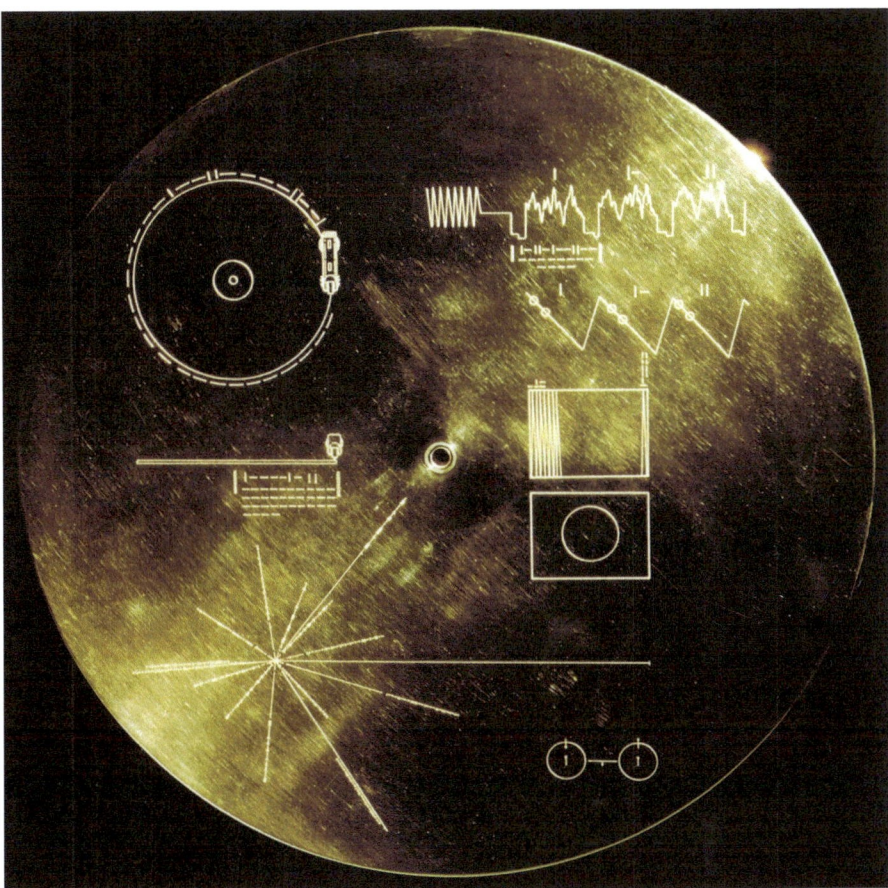

Fig. 2.3 The cover of the Golden Record which flew with the Voyager space probes. Inscribed on the cover are pictographic instructions on how to play the record, how to extract video information from it, and a pulsar map that allows the receiver to calculate Earth's galactic position at the time Voyager was launched. (Image credit: NASA/Jet Propulsion Laboratory (NASA/Jet Propulsion Laboratory. **The Golden Record Cover**. https://voyager.jpl.nasa.gov/golden-record/golden-record-cover/))

3

Animal Communication

While SETI has yet to detect a signal from an extraterrestrial civilization, animal communication research offers an analogy of what we might expect as we attempt to understand an alien communication system. We can also use techniques from animal communication research to quantify different aspects of communication systems and to rank them in terms of their sophistication. Before we get into the details of parsing and understanding a message from another civilization, it is helpful to explore how we are attempting to communicate with "aliens" here on Earth.

This research has focused on a number of key questions, among them:

Do other animals use and understand language as we think of it?
Can we understand their communication systems, and can they understand us?
How do we differentiate between rudimentary communication and language as we use it?

While humans are unique in understanding written language, many animals have the ability to communicate, and some possess the capacity for problem-solving and abstract reasoning. We have also succeeded in partially decoding the languages for some animals.

© The Author(s), under exclusive license to Springer Nature Switzerland AG 2021
B. S. McConnell, *The Alien Communication Handbook*, Astronomers' Universe,
https://doi.org/10.1007/978-3-030-74845-6_3

What's in a Language?

Whether animals have the capacity for language, and whether we can understand their language, is a question that Con Slobodchikoff, who we will meet later in this chapter, has studied in depth with his work on prairie dogs. According to him, to qualify as a language, a communication system will need to have the following seven traits:

1. **Semantics**. Each unique word or signal pattern conveys a unique meaning or small set of possible meanings depending on the context it is used in.
2. **Arbitrariness**. The word or signal pattern is arbitrary, meaning the pattern is not related to the concept it is describing. Take the word "circle" as an example. There is nothing about the word that is inherently related to a circle.
3. **Discrete units**. Larger expressions, such as sentences, are created by grouping smaller units, such as words. Each is a discrete unit that functions as a building block for larger expressions.
4. **Displacement in time or location**. A language enables the user to refer to objects or actions that are displaced in time or location (e.g., something that will happen in the future, something that is occurring somewhere else, etc.).
5. **Productivity**. A language should enable the user to create terms to describe new objects or actions.
6. **Duality (modularity)**. A language is created by combining small units into larger ones (e.g., words within sentences).
7. **Cultural transmission**. Languages are learned, not inborn.

These criteria can be applied to any communication system, whether it is audio communication among animals, a written communication system, or even a computer programming language. These criteria help us distinguish the difference between mere signaling and language. Where signaling is more programmed or instinctual, language allows its users to exchange information that informs but does not dictate behavior and, like a tool, can be used to share information and to influence the behavior of others.

Information Theory and the Structure of Language

The English language is structured in ways that can be measured empirically. While the English alphabet has 26 letters, they are not all used with the same frequency. If you graph the frequency with which letters occur in English texts, you'll see that a few letters account for the most usage. We see similar patterns in any language that employs an alphabet.

Symbol Frequency

This relationship also appears in groupings of letters, which is known as *conditional probability*. Let's say a word starts with the letter "s." Certain letters are much more likely to follow an s than others, a relationship that we can also visualize graphically.

These patterns differ from what we would expect to see if all letters were used with equal probabilities. Conditional probability turns out to be a form of error correction that enables the reader or listener to work out what was meant if some information is lost. We see similar distributions in spoken language when we break words and phrases into individual phonemes, where certain phonemes are more likely to be grouped with others. This can be extended to the third, fourth, and nth letter combinations, which will tell us how much additional information is added with each letter in a word. In English, once you know the first four or five letters in a word, there is less uncertainty about which letters may follow to form longer words.

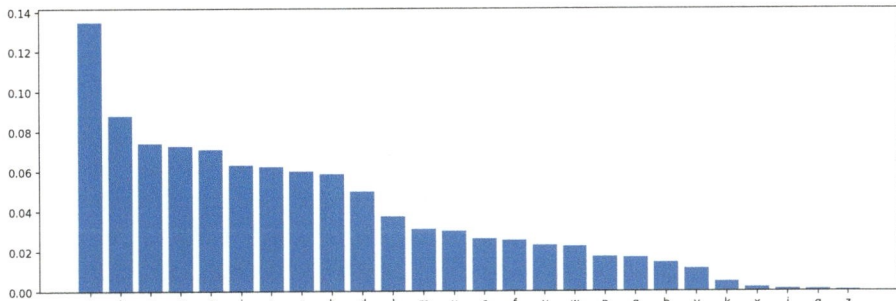

Fig. 3.1 A graph of the relative frequency of letters used in the English language. Note that the top 10 letters account for 75% of use. This pattern is seen in all alphabets. This is an example of first-order Shannon entropy, which we will discuss in more detail in Chap. 8 on Entropy. (Image credit: Brian McConnell)

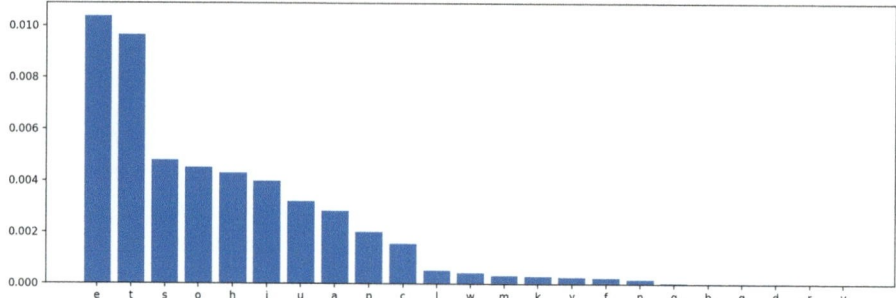

Fig. 3.2 A graph of the relative frequency of the letters following the letter s in English text. Notice how a few letters are much more likely to follow an "s" than others, an example of conditional probability or second-order Shannon entropy. (Image credit: Brian McConnell)

Higher-order probabilities also hint at structure in a language, such as patterns for grouping letters, words, or groups of words.

Zipf's Law

A similar relationship also applies to words. This is described by *Zipf's law*, which predicts that there is an inverse relationship between a word's rank and how frequently it appears. The number one word "the" accounts for nearly 7% of usage in the English language, while the number two word "of" accounts for 3.5% of usage (one half the frequency of "the"). This pattern continues all the way down to the least used words.

The Distribution of Meaning

Another pattern we see in human languages is that the most frequently used words often have multiple meanings. These words only have a decipherable meaning in the context of a larger expression. This is another example of information compression at work in language, as it enables the users of the language to compact more meaning into a small expression. The downside is that the listener needs to understand the other words used in conjunction with a word that has multiple meanings in order to decode the expression.

Let's consider the English word "like" as an example. This word is short and frequently used and, depending on its context, has many different meanings, among them:

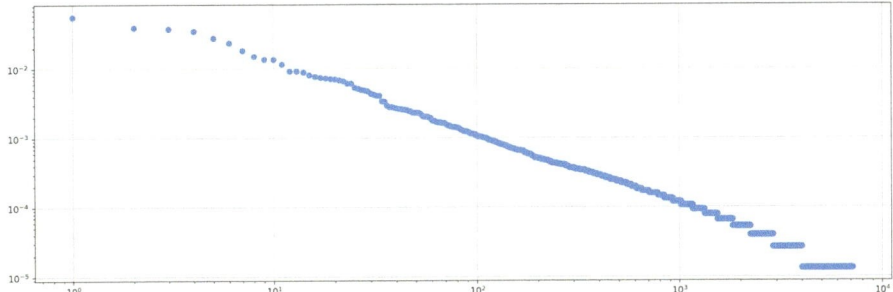

Fig. 3.3 A logarithmic graph of word frequency versus rank for Mary Shelley's *Frankenstein*. Notice that the frequency of a word's use (*y*-axis) is inversely proportional to its rank (*x*-axis). This type of power law relationship is seen in all human languages and some animal communication systems, such as bottlenose dolphins. (Image credit: Brian McConnell. Text courtesy of Project Gutenberg. (Shelley, Mary Wollstonecraft. *Frankenstein; Or, The Modern Prometheus*. (1818)))

Is similar to: "This car is like the other one we looked at"
Is fond of or has an affinity for: "Jane likes Robert"
Is the same type of object: "Put like with like"
A noun phrase: "He gave the post a like"
Placeholder statement: "Like, whatever…"

This pattern makes sense because the speaker of a natural language is limited in how fast they can generate utterances or, in the case of a written language, how fast they can write. Shortcuts like this enable the users of a language to convey more information with fewer symbols or utterances.

The Law of Abbreviation

We see a similar pattern that the most frequently used words in a language tend to be shorter in written form or require fewer syllables in spoken form. This is another form of compression that enables the users of a language to minimize the number of utterances or written elements needed to convey something.

Data compression algorithms on computers work in very much the same way. These programs scan a large block of data to look for repeating patterns and then rank them by how often each pattern occurs. The most frequently repeating patterns are assigned the shortest codewords, while less frequently occurring patterns are linked to longer codewords. The result is a compressed block of data that can be significantly smaller than the original, especially if there is a lot of repetition in the original data set. This process can be reversed on the receiving end with no loss of information.

This is also something that has been noted in animal communication research, especially in relation to alarm calls. Animals use alarm calls to alert each other to the presence of predators and, in cases of imminent danger, tend to use terse calls that compact several pieces of information into a brief utterance. These terse calls can combine information about the type of predator, the species of predator, and speed or degree of urgency. While they may appear to be a single "word," they can also convey adjective-like and verb-like information.

Recursion

The structure of the language and how components such as words and letters are combined can also tell us something about the intelligence of the language's users. Recursion is another measure of a communication systems complexity, which is illustrated in the set of examples below. We can measure the amount of recursion in a language and from that draw inferences about the sophistication of its user:

> Hand me the box of nails (r = 0)
> Hand me the box of nails <u>that Dan bought</u> (r = 1)
> Hand me the box of nails <u>that Dan bought</u> *<u>that were on sale last week</u>* (r = 2)

In the example sentences above, we can see how recursion works in human language. The first sentence is a simple statement with no recursion. The second sentence adds a relative phrase "that Dan bought" to further identify which box of nails is being referred to. The third sentence adds an additional relative phrase to identify the specific box of nails that Dan bought. This type of pattern enables the speaker to pack a lot of information into a single expression and also to rearrange a finite number of words or symbols in an infinite number of ways.

The amount of recursion in an expression can also hint at the intelligence of the language's users, because each level of recursion requires more working memory to keep track of all of the relative or dependent segments of an expression.

These patterns can be measured empirically in animal communication systems and will provide insight into the sophistication of these systems.

Together, these patterns enable the user of a communication system to transmit more information and to do so more reliably.

If we encounter an alien message, we will be able to use the same techniques to analyze its contents. Even if we are unable to understand its meaning at first, we will be able to gain insights into how it is structured and how sophisticated the communication appears to be.

Comparing Animal Communication Systems

The first point of comparison between animal communication systems is to look at the number of different utterances or behaviors an animal can use to communicate. This will give an idea of the size of an animal's "alphabet" from which larger expressions can be created.

This measure alone does not reveal much about the sophistication of a communication system, since it just measures the number of basic symbols and not how they can be combined in larger groupings to form the equivalent of words and sentences. To get a better understanding of how complex the communication system is, we will want to study higher-order entropy to see if conditional probability is at work (much more on entropy in Chap. 8).

By studying higher-order entropy in a system, one can get an idea of how complex the system is, as well as the size of its vocabulary. This does not require one to understand the meaning of individual symbols or utterances in the system, but simply involves comparing the frequency with which they and

Table 3.1 A comparison of how many different utterances or symbols are used in selected animal communication systems

Species	H_0 (bits)	Repertoire size
Grasshopper	3.58	12
Hermit crabs	4.09	17
Mantis shrimp	3.32	10
Rhesus monkey	6.91	120
Cardinal	3.32–3.91	10–15
Bottlenose dolphin	6.67	102
Homo sapiens (Hawaiian alphabet)	3.46	11
Homo sapiens (English alphabet)	4.75	27
Homo sapiens (English phonemes)	5.46	44

A comparison of the number of different utterances or behavioral signals used in animal communication systems, also referred to as zero-order entropy or H_0. (Hanser, Sean, McCowan, Brenda, Doyle, Laurence. **Information Theory as a Comparative Measure of Animal Communication Complexity**. A New Era In Bioastronomy. ASP Conference Series, Vol 213, 2000)

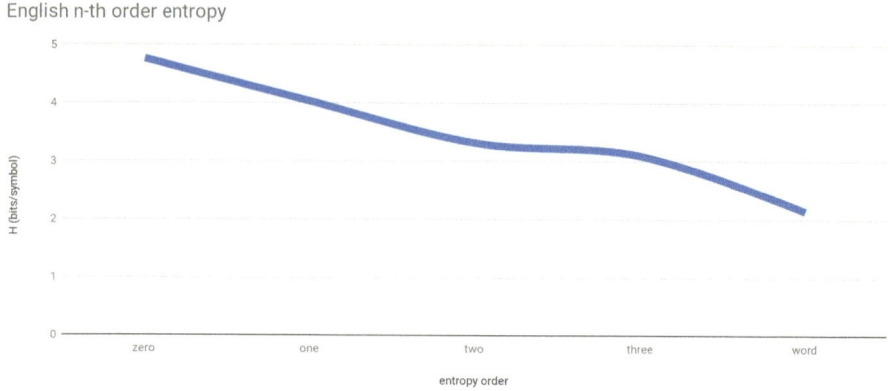

English n-th order entropy

Fig. 3.4 The graph above plots H, the number of bits encoded per letter in English text, from zero-order and onward. The number of bits encoded per letter drops for higher orders, which means that as letters are combined in larger groupings, there are fewer possible words that can be created from them and thus less uncertainty and less information with each added symbol. (C. E. Shannon. *Prediction and entropy of printed English.* Bell System Technical Journal., pp. 50–64, Jan. 1951)

their combinations are used. This also enables us to compare animal communication systems in a quantitative way by studying how entropy drops off for higher-order combinations of symbols or by measuring the amount of recursion in use Hanser et al 2004.

One of the challenges faced in animal communication is the difficulty of collecting data. It is easy to study written human languages using statistical methods, as very large data sets are readily available in digitized form, which can be fed into computer programs that do this sort of analysis. In contrast, collecting and labeling recordings of animal utterances and behavior is painstaking work.

Selected Examples of Animal Communication

Communication among animals, from insects to primates, is ubiquitous and is deeply linked to an animal's ability to survive. This communication takes many forms and ranges from chemical signaling in the case of ants to complex verbal communication among mammals. While communication is near universal among animals, this should not be confused with language in the sense that humans use it.

Here, it is important to make a distinction between signaling and language. Language, as discussed, is modular and extensible and is also transmitted via culture or learning. Many animals have sophisticated signaling mechanisms, but most of these do not pass the tests for language, as they are more programmed or instinctual.

Ants

Ants communicate via chemicals, most notably by using scent trails as they discover food resources. While this is a fairly simple mechanism, it often leads to complex or emergent behavior. As an ant traverses a path, it leaves scents that other ants can detect and follow. The more ants follow a path, the stronger the pheromone signal becomes.

This communication system is simple, but it is effective since ant colonies host a large number of individuals. While each individual may initially wander off in a random direction in search of food, as a collective, they will quickly discover nearby resources and, by traveling back and forth to them, will signal to the rest of their colony which paths are the most promising.

To an observer looking down on an ant colony, it might appear that there is elaborate and coordinated planning and communication going on. While it may seem that way, this type of emergent behavior is often seen in systems where there are a large number of agents governed by simple rules.

Bees

An interesting example of insect communication is the waggle dance, which is used by honeybees to communicate the direction and distance to nearby food resources. In this dance, the bee's angle relative to the sun indicates which direction to fly, while the duration of the waggle part of the bee's maneuver indicates how far away the resource is. This dance enables bees to communicate where food resources are located relative to the hive and with a high degree of precision.

While this dance enables bees to share information about where food resources are located (displacement), it does not pass several other tests for language, such as the ability to create new symbols (productivity) or cultural transmission (learning).

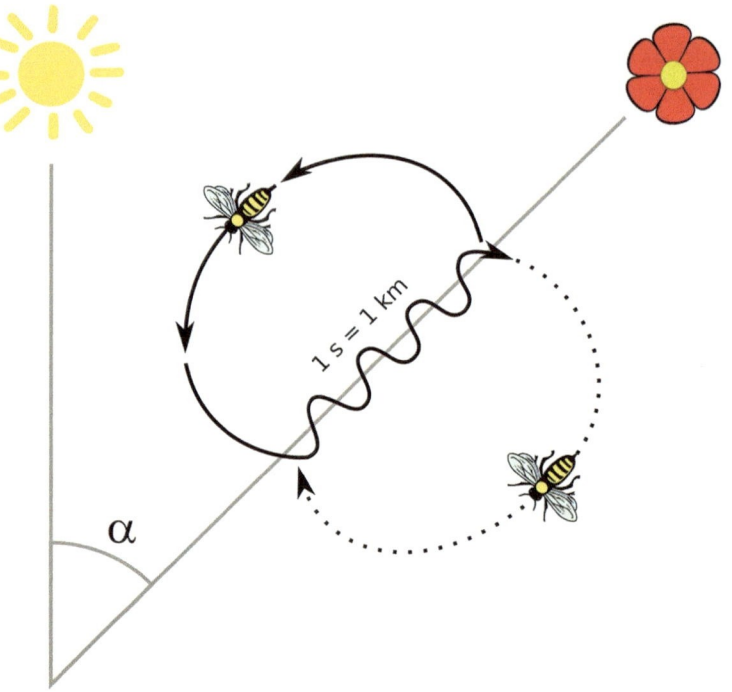

Fig. 3.5 A diagram of a bee waggle dance. (Image credit: Wikimedia Commons)

Prairie Dogs

While most readers may have heard about animal communication research with primates and marine mammals, they may be surprised to learn that prairie dogs have a sophisticated communication system that meets most of the criteria for language.

Prairie dogs are easier to study than most animals due to their "urban" lifestyle, as they live in large networks of burrows in colonies, often referred to as towns. They stay in or near their colonies, which are often adjacent to human settlements, and as a result are easier to observe than animals that reside in remote areas or roam over large territories.

Con Slobodchikoff, whom we met at the beginning of this chapter, has found that prairie dogs have a sophisticated system of calls they use to identify predators and potential threats to their colonies. In the course of studying them, he and his team found that their communication meets many of the tests for language and were able to decode parts of it.

Slobodchikoff was able to decode prairie dog calls by recording them as different predators approached their colonies and by observing how they

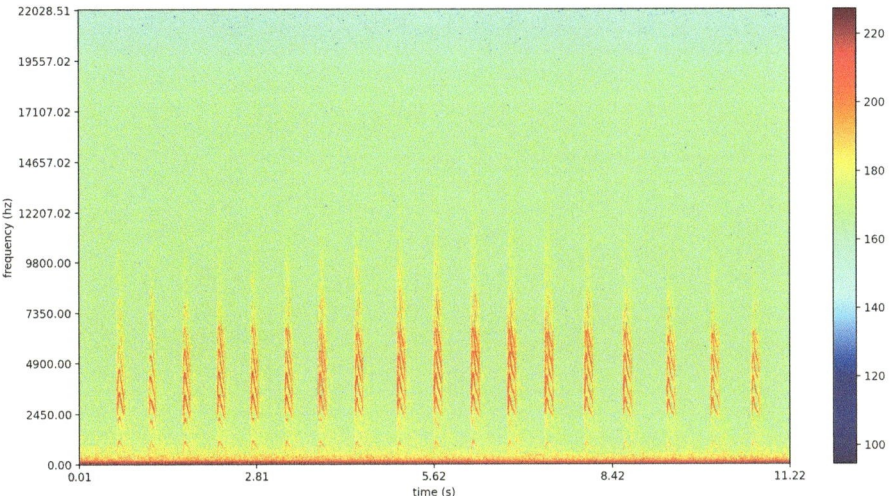

Fig. 3.6 A spectrograph of prairie dog chirps to alert others about the presence of a dog. (Image credit: Brian McConnell. Source audio courtesy of Con Slobodchikoff)

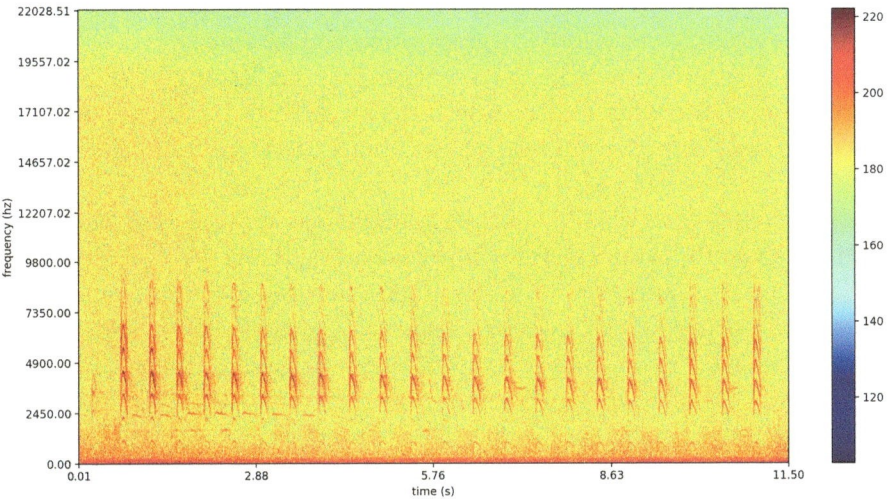

Fig. 3.7 A spectrograph of prairie dog chirps to alert others about the presence of a coyote. (Image credit: Brian McConnell. Source audio courtesy of Con Slobodchikoff)

behaved in response. Using a solve for x pattern, his team would vary one aspect of a scene or interaction and study how the prairie dogs responded to it, for example, by changing the color of the shirt a team member wore while walking through the colony, and then studying the prairie dogs' calls to see

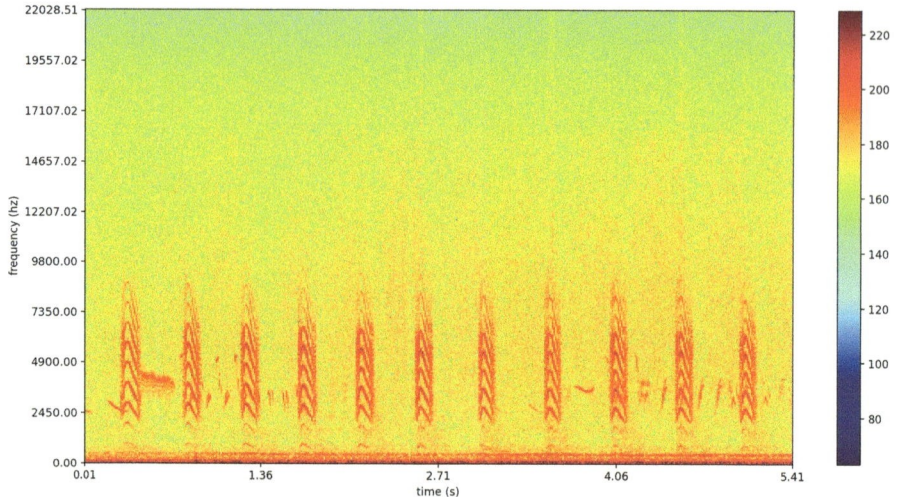

Fig. 3.8 A spectrograph of prairie dog chirps to alert others about the presence of a human. (Image credit: Brian McConnell. Source audio courtesy of Con Slobodchikoff)

what changed. This enabled the team to tease out the differences between their calls and to work out their composition and structure.

Prairie dog communication goes further than simply assigning "words" to each category of predator (they have different calls for hawks, coyotes, dogs, and humans). They also vary the calls to describe attributes of the predator. Slobodchikoff and his team found that prairie dogs could also describe them by color, height, and weight and could call out geometric shapes they had not seen before, with distinct calls for square and oval shapes[1].

The team also found that different prairie dog colonies had developed different vocabularies of calls, which suggests that the calls are at least in part a learned behavior or cultural communication. In addition, the team found that the calls functioned as sentences, where each individual chirp is a noun or an adjective that together form a more complex expression, while the rate of the chirps indicated how fast something was approaching.

Slobodchikoff's work is based on passive observations, whereas many animal communication experiments involve training captive animals to use human communication systems. For example, Koko the gorilla learned to communicate via sign language. Captive marine mammals are another example. The problem with this approach is that the interaction with humans can bias the experiments. Are the animals communicating freely, or are they replying in response to training?

---

[1] Email conversation with Con Slobodchikoff on June 22, 2020.

Con's experiments also provide insight into where we run into barriers in understanding animal communication. While he was able to work out how prairie dogs communicate about predators, other topics of communication remained opaque. One would presume that they communicate about other aspects of their environment, such as food sources. The weather is another, as their colonies are prone to flooding during heavy rains, a threat they would need to cope with. The problem is that it is difficult to tease out the response to calls about less direct concepts or threats.

How Might a Communication Engineer Approach Animal Communication?

Much of the animal communication research done to date has been based on painstaking fieldwork that makes collecting a large corpus of sample communication difficult. It is also hampered by human sensory limits and perceptual biases. The different calls made by an animal may sound similar to the untrained human ear. The process of capturing recordings and annotating them for analysis is also quite laborious, which makes it difficult to collect enough data for analysis. As a result, progress in understanding animal communication systems is often slow and incomplete.

How might things look if we approach this as a signal-processing and communication systems problem, much as we would if we were to encounter an alien signal? How might newer signal-processing and machine learning technology be applied to this task?

Artificial neural networks (ANNs), which learn to identify patterns from examples, could play a role in advancing animal-human communication in a number of areas. Such work could also inform future efforts to understand an extraterrestrial communication system should we encounter one. These machine learning tools would not magically translate animal communication, but rather could assist researchers in recording and characterizing animal calls in a more automated way.

One of the big challenges encountered in analyzing animal calls is that the underlying structure is often unclear to us. An analogy in human communication would be the experience of encountering a new spoken language for the first time. While you may clearly recognize it as human speech and even which language it is, you may have a very difficult time understanding where one word ends and the next begins, much less what any word or phrase means.

The problem with animal communication is compounded by the fact that a human listener has been trained from decades of experience to be sensitive to subtle variations of human speech and is apt to project patterns from human speech onto what they hear. One consequence is the risk that a human listener will interpret an unimportant part of an animal call as a signal, while ignoring other parts of the call that don't fit the pattern of what we are sensitive to. When you hear a crow calling, you most likely hear "CAW!," but there may be a lot of subtlety in that sound that you simply do not notice.

Artificial intelligences built on ANNs have proven themselves to be adept at recognizing patterns and are widely used in tasks such as image recognition and language translation. They can also be trained from a blank state and as such can be freed from the biases a human listener will have. When trained with a set of sounds, an ANN can learn which sounds are alike and different and can separate them out, whereas a human listener may not be able to easily pick up the differences between them.

Through such technology, researchers could build more automated recording systems that can be set up and left to run unattended for extended periods of time. A system could be designed that records continuously but only retains the recordings of the species it is designed to monitor while ignoring background noise or calls from other species. These recordings can in turn be handed off to other steps in the processing pipeline that break them down into smaller utterances and label them according to their characteristics. This will free researchers to focus more of their time on analysis and less on the grunt work of collecting recordings.

This type of automation has been employed in many fields, including astronomy. By automating the collection and classification of images, for example, these systems free astronomers to spend more time on analyzing the data they collect, rather than the mechanics of taking pictures.

Step 1: Identifying which Animal Is Vocalizing

The first step in the processing pipeline will be to identify what type of animal the system is listening to and then route that audio to a species-specific audio processing system. This is typically done using a supervised machine learning system that is given a large set of representative recordings, each of which is labeled with an identifier that maps to the species of animal making the call. This is analogous to identifying which language a person is speaking without knowing what they are saying.

Step 2: Identifying the Basic Elements of an Animal's Calls

The next challenge faced when we encounter a new communication system is to understand its general structure and its basic elements. In spoken English, there are 44 phonemes or sounds that can be grouped together to form words. These in turn can be grouped together to form a larger phrase or sentence. So, one of the hurdles in understanding an animal vocalization system is to parse a series of utterances into a smaller set of basic elements.

For each species that is tracked in this system, we would need to train a machine learning system with a large set of recordings of that animal's calls and task it with identifying the basic elements or characteristics of these calls. This is analogous to recognizing edges within an image. At this stage, the system is not trying to understand what any element means, just to identify the boundaries between utterances.

Step 3: Building a List of Possible Utterances or Phonemes

The next step of the processing pipeline will be to develop a list of the different utterances that occur within a set of samples. The algorithm that performs this task will compare each individual utterance to all of the others and score it based on its similarity or difference from them. Through this process, we can develop a list of phoneme equivalents, basic building blocks of speech. We are still not trying to understand what any one utterance means, just to identify the basic building blocks of the communication system.

Let's look at a hypothetical communication system that is based on chirp-like sounds. We notice that each chirp typically has one of three primary tones. We also notice that the chirp can vary in frequency over time, which adds three more possibilities to each: it can increase in pitch over time, stay the same in pitch, or decrease in pitch. We also notice that the attack rate, how quickly a chirp increases in volume, varies and that the sound either comes on very quickly or gradually increases in amplitude, and we notice the same thing for its decay rate, the rate at which the sound drops off. This works out to 3 times 3 times 2 times 2 or 36 possible combinations (not that far off from the 44 phonemes in human speech). The point is that a lot of information can be packed into a simple signal that may sound the same to the untrained ear.

With this in hand, we could build a sort of stenographer app that listens to an animal's calls and transcribes them into human-readable labels. The labels themselves would not mean anything, but would just serve to record and display them in a shorthand format people can read.

Let's look at our hypothetical example again. Let's suppose we want to transcribe these states into some sort of stenographic notation that a person can read easily. Each chirp can have one of the three primary tones, which we will nickname low (LO), medium (MA), and high (HI). Each tone or frequency can increase (NI), stay even/hold (HO), or decrease (DI) in time. The attack rate and decay rate for each can also be fast (FA) or slow (SO). These labels map to phonemes, so they can either be recorded in written form or sounded out.

You then end up with a series of labels like FALONIFA, SOHINISO, etc., with a total of 36 possible combinations for this example system. This example was constructed to produce labels that are easy for humans to read, pronounce, and memorize, but do not reflect the utterance's original sound or meaning in any way. We'll tackle that in subsequent steps. These labels can also be treated as unique symbols that can be analyzed statistically, for example, to study conditional entropy or look for Zipf's law patterns.

Step 4: Measuring Frequency of Use

One of the next things we'll want to do is measure the frequency of use for each basic unit and also look to see if conditional probability is at work, where one signal state is much more likely to be used after another. We can compare these metrics to other animal communication systems and to human language to get a sense of how simple or sophisticated the communication system is.

Let's look at our hypothetical chirp system again. The output from the chirp discriminator/stenographer will be a series of mnemonic labels (e.g., SOHINISO, FAMADISO, etc.) that describe the properties of each chirp within a longer sequence of chirps. With this data in hand, we can proceed to count how often each label occurs with respect to others.

The first step is to build a conditional probability map that shows the probability that a specific label will occur after another. This is similar to measuring the frequency with which English letters are used with respect to each other (an "s" is much more likely to be followed by a "c" than an "f"). We can compare these probability maps to other animal communication systems to get a sense of how they stack up.

We can also analyze these signal trains to look for the equivalent of words made by combining many individual utterances. In our example system, there are 36 basic utterances that can be combined to form the equivalent of words and phrases. Even a simple system can accommodate a large vocabulary if the utterances can be combined. Combining 1 or 2 utterances to form words yields up to 1332 unique possibilities, while combining 1, 2, or 3 utterances yields up to 47,988 possibilities. That's enough to support a substantial vocabulary even if most of the combinations are unused.

If something like Zipf's law is at play, we will expect to see certain combinations occurring a lot more often than others, while other combinations basically never happen. As we've discussed, Zipf's law predicts that a word's frequency of use is inversely related to its rank in terms of usage.

The existence of a distribution like this isn't itself a signal that communication is occurring, but it will point to the most frequently used combinations of utterances. At this stage, we still won't know what a particular utterance means, but we will be able to rank them by how frequently they appear and then start working down the list in ranked order to understand what they mean, if anything.

Step 5: Assigning Meaning to Words and Phrases

Once we understand the basic building blocks of an animal's communication system, we can then start assigning meanings to the more frequently used calls and the most frequently observed behaviors. Most animals are understandably concerned with not being eaten by their predators, so it makes sense that the more frequently encountered signals may be related to predators and alerting others to their presence. These calls and the escape responses to them are also easier to observe compared to other behaviors, so they tend to be among the better understood aspects of animal communication.

The problem is that other behaviors may be much more difficult to observe or might not be recognizable as such. Without a clear connection between an utterance and a behavior or response, it will be difficult to work out what the utterance means, if anything. As an example, birds might have a whole vocabulary around wind patterns and the weather, as that has a direct bearing on flying conditions. We could easily miss that sort of communication because it would be difficult to set up a controlled experiment that would detect it.

Table 3.2 depicts the usage of all possible English bigrams or two-letter combinations. The graph was generated by counting the number of times each bigram occurred in a corpus of English text. The larger the red bar associated with the bigram, the more often it is used. You can see right away that the vast majority of combinations are rarely used and that a short list of bigrams accounts for the majority of use.

Table 3.2 An index of English bigrams and their frequency of use

AA	BA	CA	DA	EA	FA	GA	HA	IA	JA	KA	LA	MA	NA	OA	PA	QA	RA	SA	TA	UA	VA	WA	XA	YA	ZA
AB	BB	CB	DB	EB	FB	GB	HB	IB	JB	KB	LB	MB	NB	OB	PB	QB	RB	SB	TB	UB	VB	WB	XB	YB	ZB
AC	BC	CC	DC	EC	FC	GC	HC	IC	JC	KC	LC	MC	NC	OC	PC	QC	RC	SC	TC	UC	VC	WC	XC	YC	ZC
AD	BD	CD	DD	ED	FD	GD	HD	ID	JD	KD	LD	MD	ND	OD	PD	QD	RD	SD	TD	UD	VD	WD	XD	YD	ZD
AE	BE	CE	DE	EE	FE	GE	HE	IE	JE	KE	LE	ME	NE	OE	PE	QE	RE	SE	TE	UE	VE	WE	XE	YE	ZE
AF	BF	CF	DF	EF	FF	GF	HF	IF	JF	KF	LF	MF	NF	OF	PF	QF	RF	SF	TF	UF	VF	WF	XF	YF	ZF
AG	BG	CG	DG	EG	FG	GG	HG	IG	JG	KG	LG	MG	NG	OG	PG	QG	RG	SG	TG	UG	VG	WG	XG	YG	ZG
AH	BH	CH	DH	EH	FH	GH	HH	IH	JH	KH	LH	MH	NH	OH	PH	QH	RH	SH	TH	UH	VH	WH	XH	YH	ZH
AI	BI	CI	DI	EI	FI	GI	HI	II	JI	KI	LI	MI	NI	OI	PI	QI	RI	SI	TI	UI	VI	WI	XI	YI	ZI
AJ	BJ	CJ	DJ	EJ	FJ	GJ	HJ	IJ	JJ	KJ	LJ	MJ	NJ	OJ	PJ	QJ	RJ	SJ	TJ	UJ	VJ	WJ	XJ	YJ	ZJ
AK	BK	CK	DK	EK	FK	GK	HK	IK	JK	KK	LK	MK	NK	OK	PK	QK	RK	SK	TK	UK	VK	WK	XK	YK	ZK
AL	BL	CL	DL	EL	FL	GL	HL	IL	JL	KL	LL	ML	NL	OL	PL	QL	RL	SL	TL	UL	VL	WL	XL	YL	ZL
AM	BM	CM	DM	EM	FM	GM	HM	IM	JM	KM	LM	MM	NM	OM	PM	QM	RM	SM	TM	UM	VM	WM	XM	YM	ZM
AN	BN	CN	DN	EN	FN	GN	HN	IN	JN	KN	LN	MN	NN	ON	PN	QN	RN	SN	TN	UN	VN	WN	XN	YN	ZN
AO	BO	CO	DO	EO	FO	GO	HO	IO	JO	KO	LO	MO	NO	OO	PO	QO	RO	SO	TO	UO	VO	WO	XO	YO	ZO
AP	BP	CP	DP	EP	FP	GP	HP	IP	JP	KP	LP	MP	NP	OP	PP	QP	RP	SP	TP	UP	VP	WP	XP	YP	ZP
AQ	BQ	CQ	DQ	EQ	FQ	GQ	HQ	IQ	JQ	KQ	LQ	MQ	NQ	OQ	PQ	QQ	RQ	SQ	TQ	UQ	VQ	WQ	XQ	YQ	ZQ
AR	BR	CR	DR	ER	FR	GR	HR	IR	JR	KR	LR	MR	NR	OR	PR	QR	RR	SR	TR	UR	VR	WR	XR	YR	ZR
AS	BS	CS	DS	ES	FS	GS	HS	IS	JS	KS	LS	MS	NS	OS	PS	QS	RS	SS	TS	US	VS	WS	XS	YS	ZS
AT	BT	CT	DT	ET	FT	GT	HT	IT	JT	KT	LT	MT	NT	OT	PT	QT	RT	ST	TT	UT	VT	WT	XT	YT	ZT
AU	BU	CU	DU	EU	FU	GU	HU	IU	JU	KU	LU	MU	NU	OU	PU	QU	RU	SU	TU	UU	VU	WU	XU	YU	ZU
AV	BV	CV	DV	EV	FV	GV	HV	IV	JV	KV	LV	MV	NV	OV	PV	QV	RV	SV	TV	UV	VV	WV	XV	YV	ZV
AW	BW	CW	DW	EW	FW	GW	HW	IW	JW	KW	LW	MW	NW	OW	PW	QW	RW	SW	TW	UW	VW	WW	XW	YW	ZW
AX	BX	CX	DX	EX	FX	GX	HX	IX	JX	KX	LX	MX	NX	OX	PX	QX	RX	SX	TX	UX	VX	WX	XX	YX	ZX
AY	BY	CY	DY	EY	FY	GY	HY	IY	JY	KY	LY	MY	NY	OY	PY	QY	RY	SY	TY	UY	VY	WY	XY	YY	ZY
AZ	BZ	CZ	DZ	EZ	FZ	GZ	HZ	IZ	JZ	KZ	LZ	MZ	NZ	OZ	PZ	QZ	RZ	SZ	TZ	UZ	VZ	WZ	XZ	YZ	ZZ

Peter Norvig personal website, Mayzner Revisited, https://norvig.com/mayzner.html
Source Peter Norvig, "Mayzner Revisited", *https://norvig.com/mayzner.html*

Revisiting our hypothetical animal communication system, we could generate a similar map of utterances and their combinations, and from that can work out which ones are used most frequently, and which ones have known meanings associated with them. This will also serve as a map of unknowns, directing us toward the utterances that are unknown in meaning or context but are frequently used. The unknowns are important because they might point you to new areas of communication that have so far gone unnoticed. This would allow researchers to take a two-pronged approach in parsing meaning, looking both at frequently observed behaviors and at frequently observed utterances.

Step 6: Building a Recognizer or Crude Translator

This step in the processing pipeline can feed into a recognizer or crude translator – something that could run as a smartphone app – that in addition to

Solving for X

Discerning meaning for a call or symbol can often be done through a solve for x process. This is what Con Slobodchikoff did in his research with prairie dogs. In their experiments they watched for the difference in calls in response to a single variable being changed and were able to work out the meaning of individual calls.

In one set of experiments, Con had volunteers walk through the prairie dog colony, each time wearing a different color shirt. He also had different volunteers walk through – some short, some tall, some thin, and some heavy. The prairie dogs were able to call out these attributes, and by using this solve for x pattern, Con and his team were able to figure out not only the prairie dog words for major types of predators but also the adjectives used to describe them.

The downside of this approach is that it is limited to animals that tend to stay in one place and that can be observed consistently in a controlled manner. Prairie dogs turned out to be perfect test subjects in this respect because they live in colonies and don't venture far from their burrows, so they could be observed consistently over long periods of time. This is much harder to do with animals that move around and have large territories. So it could be that there is nothing all that remarkable about the prairie dog's ability to communicate, but rather that we had the right set of observing conditions that allowed us to work out the details of their communication system.

identifying the animal species making a call could also provide some information about the call's meaning. Even if nothing approaching a conversation is possible, a technology like this would provide users with a new view of natural environments.

Imagine walking through a forest with an app that listens to the environment and tells you which animals are making calls and what they are signaling. One of the first things you might notice is they are all calling out a potential threat (you!).

Step 7: Interactive Experiments

If we succeed in building systems that can discriminate between different animal calls in real time, we could design experiments that evaluate their ability to use their speech to direct outcomes.

One class of experiment would test animals on their ability to use vocal utterances to control a machine. The general design pattern would be to pass the output from an utterance/phoneme recognizer into an interface that controls a machine. This could be as simple as a machine that dispenses different treats, or it could be something more complex. These experiments would test animals' ability to associate utterances with control inputs to a machine and from there go on to more complex experiments to test them on problem-solving, communication, and abstract reasoning.

This class of experiment can also get at whether an animal's communication is intentional or just part of an instinctive signaling mechanism. If an animal can learn to control a machine via arbitrary utterances, that will be pretty good evidence that they can use speech as a tool to produce a desired outcome.

Step 8: Two-Way Communication Experiments

The same processing pipeline could be used in experiments to test the limits of two-way communication. These tests would evaluate an animal's ability to pass the tests associated with language, such as the ability to associate new utterances with new objects or concepts, vocabulary capacity, cultural transmission, and other criteria.

Experiments based on a process like this would also have the benefit of using animals' existing communication system versus imposing a human-mediated system on them, as has often been the case in primate and marine mammal communication research, particularly with captive animals. The key difference is that instead of us trying to coax animals to respond to human communication or cues, we would be using technology to assist us in communicating on their terms.

Applying these Techniques to SETI

Readers might ask: What do birds possibly have in common with an advanced alien intelligence? How could our attempts to understand birds inform our attempts to understand alien communication?

When we first detect an *information-bearing signal*, meaning a signal that is modulated in some way to encode information, we will know absolutely nothing about how it is structured, what that structure might mean, or what type of information it is transmitting. It will be a lot like listening to a group of birds in the forest. It may be clear that there is some sort of signaling going on, but we will be stumped about what is actually being signaled.

So where do we even start with this?

The processing pipeline for analyzing alien communication will be similar to how we could approach the challenge of analyzing animal communication, especially in the early stages.

First, we will want to understand the basic modes of communication. If the signal is a radio transmission, it may be changing frequency over time to encode information. This is a lot like how a bird changes the pitch of its chirp over time. At this stage of analysis, we won't be trying to understand what anything means, just which forms of modulation are in use.

The second step will be to understand how the signal's modulation is used to communicate basic units of information. We might find that each carrier in the alien signal hops between several different frequencies. Each frequency can encode a different symbol, so we can work out that each chirp translates to one of several symbol states. Here we want to understand how many basic units of information, akin to phonemes, are in use. This is similar to the step of figuring out how many different utterances an animal can generate.

The next step will be to understand how often symbols are used relative to each other and if there is a system for combining them into groups or larger structures. How simple or elaborate those structures are will tell us a lot about the sophistication of the system as well as the types of information it may be delivering. Here we will use a variety of techniques, such as entropy analysis, to study how the communication stream is structured.

If we encounter language in an extraterrestrial transmission, it is likely to be an artificial or constructed language versus animal sounds or utterances. Think of this as a numeric language, where each unique concept is associated with a number. We will be able to use similar statistical techniques to parse this language, to identify the most frequently used "words" and the concepts that are most interlinked with others, and zero in on these for further investigation.

The key difference, and where the analogy starts to break down, is that a digital communication channel can interleave many different types of information, some of which may be things like digitized images and some which are more like language. Another difference is the message may be didactic, in which case the sender will be likely to cross-reference different representations of what they are trying to communicate, for example, by linking images to symbols that represent a particular idea or concept to be taught.

The Limitations of Animal Communication as a SETI Analog

Animal communication research provides an example of how the post-detection analysis and comprehension effort might proceed, but it has some important limitations as an analogy, especially as we get further along in the process of analyzing the transmission. An engineered signal from an ET civilization will be different from natural communication systems in several ways.

Interactivity

The important difference between animal communication and an interstellar communication link is that the latter is likely to be an engineered system designed to satisfy the requirements for reliably communicating across interstellar distances and may also be designed to carry a variety of media types or modes of communication. This communication link is best thought of as a piece of infrastructure that can convey many types of information, an extreme case of a wireless communication network. In other words, the medium will probably not be the message, but rather a pipe through which messages can be sent.

The multi-year round trip times imposed by the speed of light, even for the closest solar systems, will preclude anything resembling a real-time conversation or interaction. One notable exception to this is if the transmission includes computer programs designed to interact with the recipient, which we will discuss further on in the book.

The lack of real-time communication between the two parties means that the communication system should be designed so that the recipient can learn to parse the transmitted information with little or no assistance from the transmitter. This is quite different from animal communication, where interaction and learning is two-way and occurs in real time.

Intent

An engineered communication system designed to operate at interstellar distances also differs from natural communication systems in that the sender is attempting to communicate with or at least transfer information to a counter-party who has no prior knowledge of the communication system. A well-designed system will provide the receiver with multiple paths toward comprehending the information sent and may also rely on presumed common knowledge rooted in math or science.

If the sender is transmitting the information with the intention of initiating contact, it may be didactic in nature. Our attempts to understand animal communication are more akin to eavesdropping on someone else's conversation and attempting to infer what is being discussed by watching how each party behaves in response to the other.

Constructed Languages (Conlangs)

An engineered transmission may also differ from natural communication systems by employing a constructed language. The transmission might be based in part upon *semantic networks*. A semantic network is a graph of concepts, each of which is assigned a unique numeric address, similar to a telephone number or IP address. Think of this as an address space for ideas. The numeric address itself means nothing and is just used to uniquely identify each concept in the network. The receiver can learn how concepts are related to each other, for example, in terms of set membership. This type of language differs from natural languages as it is an engineered system and can also be designed to be parsed by a computer program.

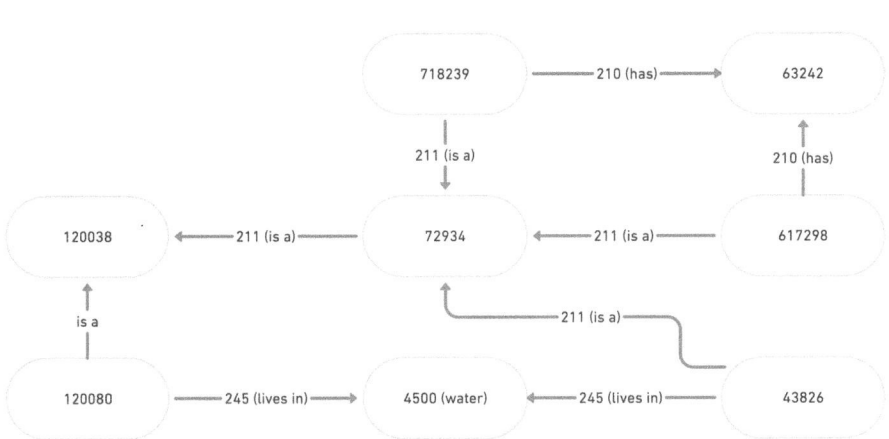

Fig. 3.9 An example of a partially decoded semantic network. Each unique concept or symbol has a unique numeric address that prevents it from being confused with others, while other symbols or operators describe how different ideas in the network are connected to each other

Let's consider the concept of a flame. This is a difficult concept to explain only in words without referencing a real-life occurrence of a flame. A sender who is trying to communicate this concept to a naive recipient could send several representations of a flame and link these to an address in a semantic network that is associated with this concept. The sender might include several representations of a flame, such as a still image, a sequence of images, or a computer program or simulation that the user can interact with.

Each of these representations could then be linked to the numeric address associated with the concept of a flame. The recipient then has multiple paths toward understanding this concept, which once learned can be mapped to others in the semantic network.

While it would be very difficult for us to understand an alien species' natural language, an engineered language, a sort of machine readable version of Esperanto, may be more straightforward to analyze and understand, at least in part.

Esperanto and Constructed Languages

Esperanto, created by L. L. Zamenhof in 1887, is the best known example of a constructed language. The language was created as a bridge language with the goal of fostering international communication and world peace. It was designed to be easy to learn and used simplified conventions for defining word usage, verb tenses, and other elements that vary widely across different languages. Most languages are riddled with exceptions for how words and phrases are formed and combined, and as such, it typically takes learners years to master them. Esperanto was designed so that someone could quickly learn the basic rules or grammar of the language and rapidly become proficient.

Alas, Esperanto did not achieve a critical mass of usage, though it remains a niche language that is used to this day. It is estimated that there are on the order of a million people who are proficient in the language. It is interesting to ponder how the language might have evolved if English had not become the lingua franca during the early development of the Internet. There was clearly a need for a bridge language as the Internet developed, and had English not been widely enough spoken, who knows, perhaps Esperanto would be the world's de facto language of trade and international communication today.

References

Vakoch, D. & J. Punske. (eds.). Xenolinguistics: Toward a Science of Extraterrestrial Language.

Hanser, Sean, Doyle, Laurance, McCowan, Brenda, Jenkins, Jon. *Information Theory Applied to Animal Communication Systems and Its Possible Applications to SETI*. Bioastronomy 2002: Life Among The Stars IAU Symposium, Vol 213, 2004.

4

A Timeline of Events

What would the timeline of events look like if we detect evidence of another civilization? This will depend a lot on the type of discovery and whether we receive information as a result of the detection or contact.

Signal or Technosignature Detection

The initial detection of an ET signal or a technosignature could happen at any time, as there are a number of surveys that are running more or less continuously. Whether the candidate signal will be quickly recognized as alien is another matter.

If the signal is clearly modulated to encode information, it may be recognized as an alien signal within a matter of days to a few weeks. A pulsed laser or optical signal is an example of something that will clearly stand out as artificial. It will also be easy to recognize variation in pulse timing as a form of modulation, so in a case like that we will know quickly if it is a bona fide transmission from aliens.

In the case of an optical signal, we could know within days if the signal is modulated and start extracting data from it Video conversation (2020). A radio signal may take somewhat longer to confirm, primarily to rule out human-generated interference, and it may take longer still to extract modulated data depending on how the signal is architected. If we were to discover inscribed matter probes, it may take months or years to recover the probes and read data out from them depending on where they are located.

B. S. McConnell, *The Alien Communication Handbook*, Astronomers' Universe, https://doi.org/10.1007/978-3-030-74845-6_4

If the signal is a technosignature, such as a chemical fingerprint in an exoplanet's atmosphere, we might recognize the possibility that it is caused by an alien civilization fairly quickly, but there would be a years' long and vigorous debate to rule out explanations that do not invoke aliens.

Depending on the nature of the signal we detect, this phase of the process could last anywhere from days to years.

Signal Analysis and Demodulation

The first question will be whether the transmission or inscribed matter is modulated or designed to encode information. This is important because it will indicate whether the signal or artifact is designed to convey information, in which case, it might be an intentional attempt to establish communication. It is also possible that we would encounter what's known as *leakage radiation*, such as a powerful radar that is used to map celestial objects such as asteroids and comets. This type of signal is not typically designed to transmit information, so if we happened to be in its line of sight, we could detect it, but we would not be able to extract any information from it. All we would know is that an artificial signal is present.

A modulated signal is modified in a way that allows it to encode information. A simple example is a radio transmission that jumps between two distinct frequencies or tones to represent a binary code. One of the first things astronomers will do following a detection event is to closely study the structure of the signal and how it varies over time, looking for signs that it is modulated in some way, and if so what type or types of modulation are employed. From there, we will be able to begin the process of transcribing data from the signal so that others can work on figuring out what information is being sent and what it represents.

This will be the most critical post-detection work and will happen in the weeks and months following the initial detection. It is important because through it, we will quickly find out if the signal is an attempt at communication or not. If it is leakage radiation, that is a potentially dangerous scenario because we may never know what the intent behind the signal is. We will only know that extraterrestrial civilizations exist in some form. As discussed earlier, the problem here is that there will be no shortage of people who claim special knowledge about the aliens or who invent conspiracy theories about why the contents of the transmission are being kept secret.

We might discover that it is not just one signal, but a collection of many signals, some of which are readily detected by our equipment, others that are much fainter and require us to build larger telescopes to clearly see them. If we find one signal, we will probably suspect that other, weaker signals may be nearby and will likely focus our equipment and attention in searches to find them.

This process will probably unfold over a period of years, especially if we need to build larger, more sophisticated receivers to detect and extract information from lower-power carriers. If we are lucky, we will be able to extract information from higher-power carriers, while we search for lower-power carriers that transmit at higher data rates.

It is also possible we will be stymied and won't recognize a modulation scheme, either because we don't understand it or because the signal was never intended to transmit information. If that's the case, we'll be blocked and won't proceed to the next steps in this timeline.

One-Way Communication

Now let's say that we do find signs that the signal is modulated to transmit information. This will enable us to proceed from signal detection to message analysis and comprehension. While signal detection and processing work will be limited to a fairly small number of astronomers and subject matter experts, the message analysis and comprehension effort will be open to people from all fields of study. It is likely to be among the largest scientific endeavors in history because of the number and diversity of people who will be able to participate. What will that process look like?

First, astronomers and subject matter experts will figure out what modulation schemes are in use and will begin transcribing raw data from the signal. This work will be done primarily by astronomers, probably with support from experts in wireless communication, digital signal processing, and related fields. Other teams will look for signs of fainter signals designed to transmit data at higher speeds but which may be operating at lower power levels and therefore will be harder to detect. How quickly this work proceeds will depend on how clear the modulated signal is within initial observations.

It is possible that astronomers will be able to see evidence that the signal is modulated, but not be able to extract every bit of information. Imagine a strobe light that flashes several hundred times per second, which is being recorded by a camera that takes only thirty pictures per second. The camera will see the strobe light as being always on or varying slightly in brightness from frame to frame, but it will not see a clear on-off-on-off pattern, since the camera takes pictures at a frame rate slower than the strobe's pulse rate.

Integration Time

The ability to clearly discern signal states is likely to be an issue for radio-based SETI detections. Radio-based SETI searchers typically break an incoming signal out into millions of narrowband channels and then integrate these signals over time. Doing so amplifies the signal, while random background noise on each channel cancels itself out. It is common practice for SETI signal analysis programs to use integration times of one second or more.

The use of long integration times may obscure modulation if the modulation rate is greater than the integration time. In this case, the encoded data may be corrupted or blurred out entirely, similar to the strobe diagram in Fig. 4.1. In a situation like this, we may detect an artificial signal but won't be able to extract information from it until more sensitive receiving equipment becomes available.

This should be less of an issue for optical SETI detection scenarios, since those systems are geared to look for pulses that can be modulated by varying the duration or timing of the pulses, a pattern that should be easier to read out.

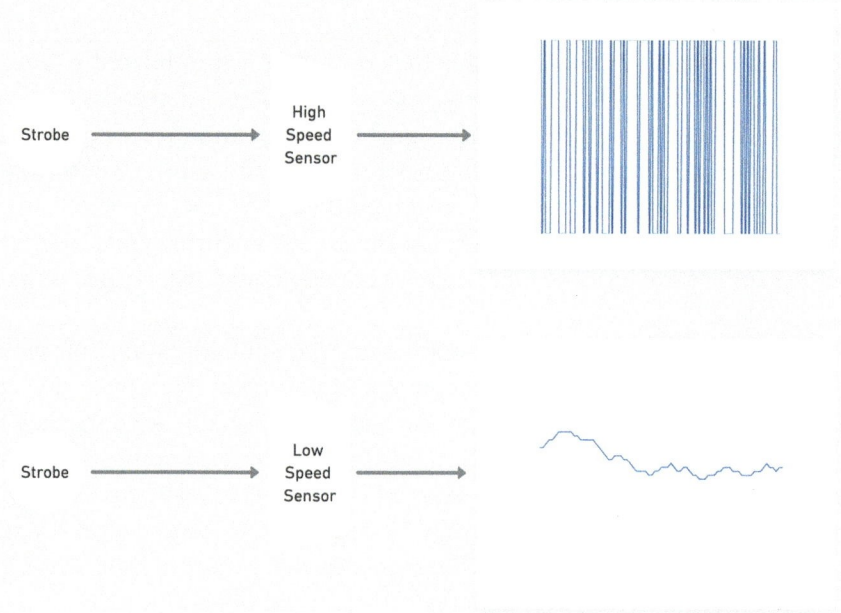

Fig. 4.1 Illustration depicting a modulated strobe light, as seen by high-speed and low-speed sensors. The output from the high-speed sensor clearly shows the strobe's signal states, while they are smeared together by the low-speed sensor, leading to loss of information

Once the basic modulation scheme is figured out, the information encoded in the signal can be transcribed into a format that analysts can work with. We won't know at this stage what the data represents. For example, if the signal is a radio carrier that chirps at one of four unique frequencies, we can describe that as a series of placeholder symbols, such as A, B, C or D. We won't really know what these placeholder symbols represent, just that each chirp codes for one of 4 unique states or 2 bits per symbol.

The raw data transcribed from the signal will be made universally available via the Internet, probably via a relay site that is designed to archive this data and to handle a large number of users. This system – we'll call it the Interstellar Communication Relay McConnell, Brian. (2020)– will appear soon after the detection event. It may be a SETI-sanctioned effort, something organized by Internet organizations and technology companies, or a distributed effort run by volunteers. Whether through the result of an organized effort or not, anyone who wants access to this data will probably have access to it.

While the relay and data archive will make it possible for amateur and professional analysts to fetch data as it arrives, other systems and journals will be created for analysts to publish and discuss their findings. It is likely that new scientific journals will be formed specifically for the analysis and discussion of the interstellar signal and its contents.

While the initial detection effort will be led by a relatively small group of astronomers, the message analysis and comprehension effort will involve amateur and professional scientists from around the world. The most useful insights in comprehending the message will most likely come from people who are not presently involved in SETI.

From there, researchers will study how this information is organized and how to parse it. This process may unfold pretty quickly if the transmission is designed to be parsed and understood by a naive receiver. That assumes the transmission is designed for us. It may just be that we happen to be eavesdropping on a transmission meant for someone else, in which case we might not be able to make any sense of it.

If the transmission is *modular*, meaning it is broken down into many smaller pieces or media types, we may be able to figure out some parts of the transmission pretty quickly and gradually work our way up the "difficulty level" to tackle more challenging information types.

It is also likely that the pace and depth of communication may ramp up over time. At the beginning, we may be limited to receiving smaller amounts of information, due to limitations in receiving equipment or to limited knowledge about how the signal is structured. At the same time, we may struggle to parse some parts of the transmission, while others, such as images, are more accessible to us.

Deciding whether to Respond

The "Declaration of Principles Concerning the Conduct of the Search for Extraterrestrial Intelligence," adopted by the SETI Permanent Study Group of the International Academy of Astronautics (1989), (2010) stipulates that any decision about whether to transmit a reply to an information-bearing message from an alien civilization should be taken up by a representative global body such as the United Nations.

This debate will likely be drawn out and vociferous, as there are many people in the SETI community who harbor strong reservations about actively messaging other civilizations. How this debate would play out will probably depend a lot on the information content of the transmission and our ability to understand it. If the meaning and intent of the transmission are unclear, contact skeptics will have a potent argument that we should be wary of responding to someone or something we do not understand. If we are successful in understanding parts of the transmission, and especially if the sender provides evidence that they are already aware of our existence, the risks of responding will be lower, as should the objections to sending a return communication.

While there would be extended debate about whether to transmit a signal in reply, it will be difficult to enforce a worldwide gag order to prevent an organization with sufficient resources from doing so. Whether or not we transmit a global response to the transmission, it is likely and perhaps inevitable that some sort of reply from at least a few rogue groups will be transmitted in the months and years following the detection of an ET signal. Indeed, organizations like this may force the rest of the world's hand.

Transmitting a Response

If we do agree on sending a global response, the challenge then will be to agree on how to organize the response and what to say. The structure and content of the alien transmission will likely inform our response. We might decide to mimic the structure of the incoming message and its contents. If the transmission contains images of organisms from other worlds, we might respond by sending a catalog of life on Earth.

Then again, we might fail to reach consensus on what to say and instead respond with a grab bag of responses from different authors in a sort of interstellar Twitter feed. The debate about how to organize the response and what to include will probably be an ongoing process since the return transmission could continue for many years or indefinitely.

Two-Way Communication Begins

Any replies we do send will take several years at a minimum to reach the transmitting civilization. While nothing resembling a real-time conversation will be possible, the structure and content of the reply will provide information to the sender about our capabilities, organization, and other characteristics. Perhaps instead of thinking about this like an instant message chat, a better metaphor would be the exchange of letters by sailing ship before the development of electronic communication systems.

Upon receiving a return transmission or transmissions from Earth, the sender would begin to learn what information we had successfully obtained, as well as what we had difficulty with, and could adapt the content of their ongoing transmission to better match our capabilities or senses. They would also be able to signal that they have received our response by echoing part of our response in their outgoing transmission.

The conversion may also evolve over time as we learn more difficult or obscure aspects of their transmission. Much of this will depend on how successful we are in comprehending what they are saying and in learning how to compose responses that are intelligible to them. This process could unfold over decades or generations, as each side learns from the other and adapts their response to maximize their intelligibility to the other.

How long would this continue? If we can understand them at least in part, communication could continue indefinitely, especially if we find ourselves in communication with a node in a larger network of civilizations. If the information we receive is interesting or is of value to us, it seems likely that communication would continue long term.

There is a scenario where meaningful two-way communication could begin much sooner in this process, and that is if the transmission contains generally intelligent algorithms, or we find ourselves in communication with an intelligent Bracewell probe located in near-Earth space. In either of these situations, the long delays imposed by the speed of light would be mooted. How sophisticated this communication ends up being will depend on the capabilities of these algorithms and our ability to understand and interact with them.

References

Video conversation with Shelley Wright at UC San Diego NIROSETI Program, Friday, Dec 4, 2020.

McConnell, Brian. (2020). **The interstellar communication relay**. *International Journal of Astrobiology*, 1–4. doi:https://doi.org/10.1017/S1473550420000178.

International Academy of Astronautics (1989), Declaration of Principles Concerning Activities Following the Detection of Extraterrestrial Intelligence.

International Academy of Astronautics (2010), Declaration of Principles Concerning the Conduct of the Search for Extraterrestrial Intelligence. *SETI Permanent Study Group of the International Academy of Astronautics*.

5

Information Delivery (Carriers)

What methods can be used to convey information across the vast distances of interstellar space? The surveys searching for evidence of communication from other civilizations look for media that are economical (do not require vast amounts of energy), can stand out against background noise or natural processes, and can be modified in some way to encode information. Based on our current knowledge of physics, three candidates stand out, two of which are thought to be the most likely conduits for extraterrestrial communication.

Electromagnetic Radiation

Electromagnetic radiation is pervasive, part of nearly every interaction involving ordinary matter. We experience it every day as light, though visible light is only a small part of the electromagnetic spectrum. SETI surveys concentrate most of their efforts looking for artificial signals that use specific parts of this spectrum.

Microwave or radio SETI primarily looks for transmissions in the microwave spectrum and is considered a branch of radio astronomy. These surveys look for radiation whose wavelength is measured in centimeters (or measured in terms of frequency, from a few hundred megahertz out to about ten gigahertz). Signals in this part of the spectrum can readily pass through the Earth's atmosphere, so they are visible to ground-based telescopes, and because the sky is relatively quiet at these wavelengths, they do not need to compete with powerful background sources, although they do have to contend with human-generated radio interference.

© The Author(s), under exclusive license to Springer Nature Switzerland AG 2021
B. S. McConnell, *The Alien Communication Handbook*, Astronomers' Universe,
https://doi.org/10.1007/978-3-030-74845-6_5

Optical SETI primarily looks for transmissions in visible light, the part of the EM spectrum we can see with our eyes, as well as in parts of the infrared spectrum. These searches look for very brief flashes of light, usually a billionth of a second or less, as no known natural processes emit such short bursts of light. A laser that compresses all of its light into a short pulse can briefly outshine the star behind it, a signal that can be detected at interstellar distances.

Inscribed Matter

Most discussion about SETI focuses on electromagnetic radiation (radio or lasers), but if a civilization wishes to send large amounts of information, inscribed matter is an attractive way to do so. *Inscribed matter* is simply matter that is altered in some way, or inscribed, to record information. Compact discs, which record digital information as pits that are etched into the surface of the disc, are a familiar if rudimentary example of the concept.

Christopher Rose, an electrical engineering professor at Rutgers University and Gregory Wright calculated how much energy would be required to transport information across interstellar distances using inscribed matter and compared this to the energy required for electromagnetic communication in a paper published in *Nature*. (Rose et al) They showed that in terms of the energy required to transmit a given amount of information, inscribed matter often wins out over electromagnetic communication, primarily because such large amounts of information can be etched into a block of metal at near atomic scales.

The biggest advantage that inscribed matter will have over electromagnetic signals is that it can be deposited over time at its destination and picked up by the recipient at any point. This differs from electromagnetic radiation, which is ephemeral and must be detected by the recipient at the instant it crosses their location.

Gravitational Radiation

With the LIGO observatory, we acquired the ability to detect *gravitational waves*, fluctuations in the fabric of space-time, and since 2015 have been observing gravitational waves generated by events such as black hole and neutron star mergers. Gravitational waves have been posited as a potential carrier for interstellar or even intergalactic communication. However, they require extraordinary amounts of energy to generate, so unless there is an as yet unknown way of generating them using less energy, electromagnetic and inscribed matter will be more attractive from an energy efficiency standpoint.

That said, we don't know what we don't know and should watch observations from LIGO for unexpected signals. LIGO, unlike radio and optical telescopes, is an omnidirectional system and can "hear" incoming gravitational waves coming from any direction at all times.

An artificially generated gravitational wave could call attention to itself by following an unexpected pattern, perhaps by maintaining a constant frequency or decreasing in frequency, neither of which would be consistent with a naturally generated signal. The cost of looking for unexpected signals is negligible, because it is primarily a matter of writing signal-processing software to scan data from the telescopes to look for other types of signals.

Besides looking for ET, there are other reasons to look for unexpected signals, as they may reveal information about as yet undiscovered natural processes. It has often been the case that whenever we developed a new way of observing the universe, we noticed new phenomena that in turn led to new scientific discoveries.

Other Carriers

Perhaps ETs have found a better medium for communication than electromagnetism or inscribed matter. Many possibilities have been suggested, but because they typically involve physics that we do not yet understand, it is not possible to design instruments to detect them. What we can do is watch for unexpected signals from experiments that were built for other purposes. Dark matter experiments, which seek to understand what dark matter is made of and how it interacts with ordinary matter, may for example, unearth an unexpected signal that turns out to be artificial.

Reference

Rose, Christopher Wright, Gregory. **Inscribed matter as an energy efficient means of communication with an extraterrestrial civilization**. *Nature* 431, 47–49, https://www.nature.com/articles/nature02884.

6

Modulation

The first thing researchers will do upon confirming the detection of an extraterrestrial signal is to look for signs that it is changing over time, which could be a clue that it is being modulated to encode information. This work will be done primarily by astronomers and signal-processing experts and will probably not involve the general public.

This chapter explains the different ways that signals can be modulated to encode information. These same methods are used in terrestrial communication systems, so readers can learn a bit about how such systems transmit information, even if they are not directly involved in the effort to understand the extraterrestrial signal structure.

Encoding information onto a radio or optical carrier is done using a process called modulation, which we have mentioned throughout this book. Let's say we want to use sound to transmit information across a room – a good metaphor for transmitting information via radio. One way to do this is to use the pitch of the sound to represent a digit. Using standard musical notes, we can represent 12 unique states within each octave. To keep the example chapter simple, we'll use binary code, so each transmitted digit can only be a zero or a one. The transmitter arbitrarily decides to use a C note (261.63 Hertz) to represent a zero, while an F note (349.23 Hertz) will represent a one.

The receiver on the other side of the room will hear a series of musical notes, always either a C or F note. The meaning or organization of this signal

© The Author(s), under exclusive license to Springer Nature Switzerland AG 2021
B. S. McConnell, *The Alien Communication Handbook*, Astronomers' Universe,
https://doi.org/10.1007/978-3-030-74845-6_6

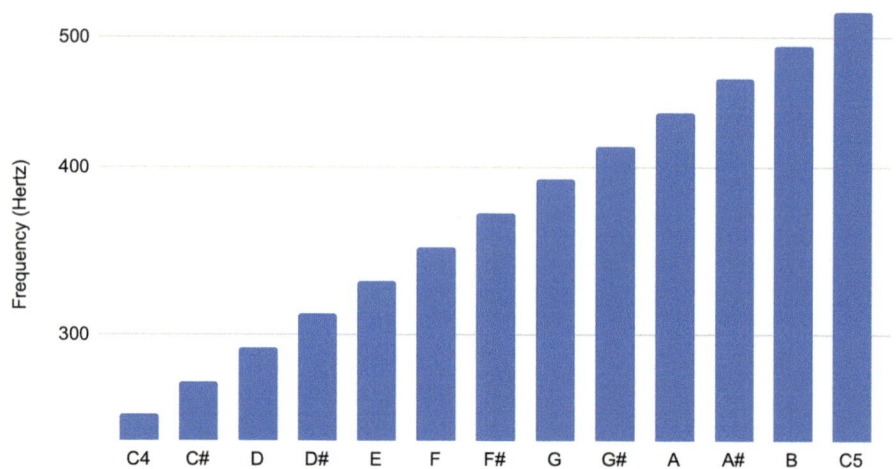

Fig. 6.1 A standard 12 note per octave musical scale. (Image credit: Brian McConnell)

won't be obvious to the listener, but it will be apparent that the sound transmitted is always one of these two notes. Anything that can be represented as a series of numbers can be transmitted via this type of system. This is a simplified example of frequency modulation, which we will discuss shortly.

An interstellar communication engineer who is tasked with designing a signal to maximize its detectability and comprehensibility will be concerned with two conflicting goals. The primary goal is to make the signal easy to see. This is done by making the signal highly monochromatic, so the transmitter's power is concentrated in a narrow band of frequencies, or by making it a very brief but very strong timed pulse. Both of these patterns are the opposite of what naturally generated radiation looks like.

The second design goal is to make it easy for the receiver to see how the signal is modulated to encode information onto it. The receiver won't immediately know what the information represents, but if they can see that the signal is a radio carrier that chirps on one of four different frequencies, they'll be able to deduce that each chirp could represent two bits (2^2 or 4 possible states) and can start testing that guess by examining data transcribed from the signal.

Radio Signals

SETI started out as a branch of radio astronomy. The first SETI survey, Project Ozma, was conducted by Dr. Frank Drake at the Green Bank Observatory in 1960. The search examined the stars Tau Ceti and Epsilon Eridani for evidence of radio transmissions. Radio-/microwave-based SETI is now a well-established field with ongoing programs headquartered at the SETI Institute, UC Berkeley SETI Research Center, Harvard, and the privately funded Breakthrough Listen initiative.

Radio signals can be modulated in a number of ways to encode information, which include amplitude modulation (AM), frequency modulation (FM), phase modulation, and polarization modulation.

Amplitude Modulation

Amplitude modulation works by modifying the strength of the signal over time. To transmit digital information, the transmitter would switch between finite numbers of transmitter power levels. This is how the first radio broadcasts worked, and it is still used for AM radio today.

Figure 6.2 depicts an AM carrier modulated with a binary code. This signal transmits at one of two power levels. In this graph, the vertical axis represents signal strength, while the horizontal axis represents time.

It is possible to reduce background noise by integrating the signal over time, which is similar to averaging multiple values, a moving average. The random noise will cancel itself out while the artificial signal is reinforced. This will blur the modulation pattern, which becomes harder to discern as the integration time is increased.

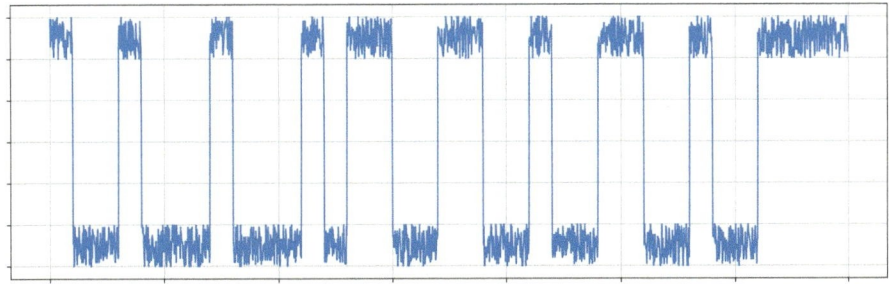

Fig. 6.2 An AM-modulated signal that hops between two distinct amplitudes with some background noise added. (Image credit: Brian McConnell)

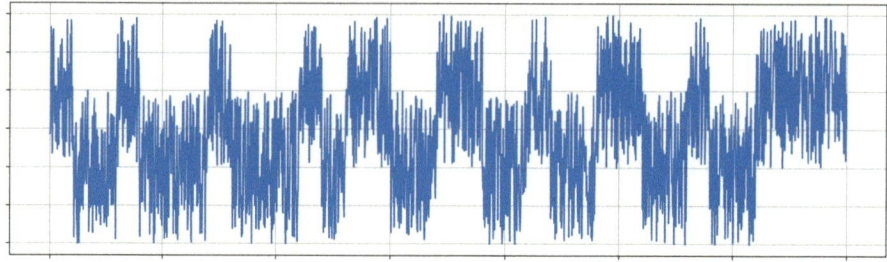

Fig. 6.3 The same AM carrier, with stronger background noise. Notice that it is diffi-cult to see the individual signal states. In a case like this, the receiver will have a diffi-cult time correctly reading out the signal state, which will result in transcription errors. (Image credit: Brian McConnell)

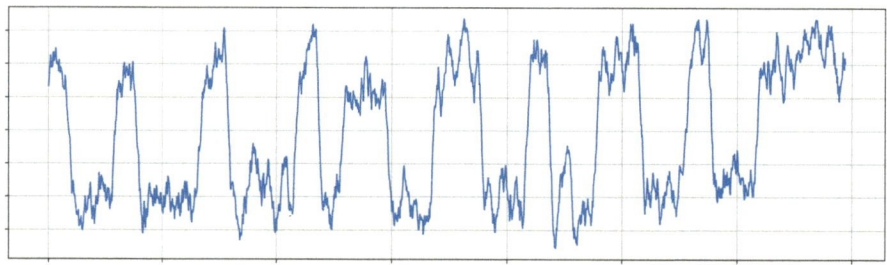

Fig. 6.4 A noisy carrier after applying a moving average calculation to it. Notice that the noise is reduced and the signal states are more clearly visible. (Image credit: Brian McConnell)

In Fig. 6.4, the noisy signal has been integrated at 15 samples per integra-tion. Notice how the noise is decreased and the modulated signal becomes more apparent.

Pulse Width Modulation

If the signal is pulsed, the duration of the pulse can be varied to encode infor-mation. Let's say that the signal chirps several times per second. The duration of each chirp can be used to encode a bit of information. For example, a 100 millisecond chirp could be used to represent a zero, while a 200 millisecond chirp could be used to represent a one. The receiver would see this by plotting the number of chirps received versus chirp duration and would see that there are two clusters, one around 100 ms and one around 200 ms. This would be a clue that some sort of pulse duration or pulse width modulation is in use.

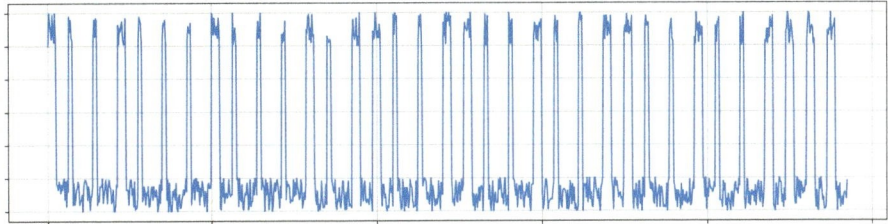

Fig. 6.5 A plot of a pulse width-modulated signal, with some background noise added in. The vertical axis represents signal strength (amplitude), while the horizontal axis represents time. (Image credit: Brian McConnell)

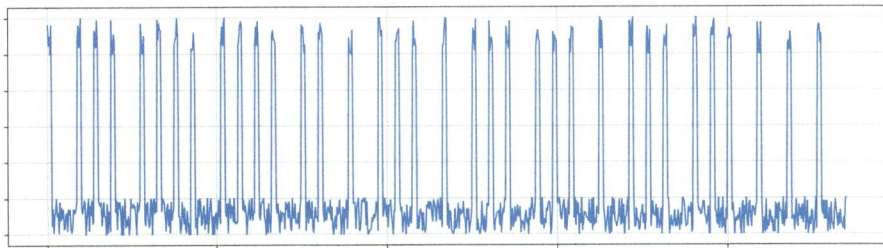

Fig. 6.6 A plot of a pulse interval-modulated signal with some background noise added in. The vertical axis represents signal strength (amplitude), while the horizontal axis represents time. (Image credit: Brian McConnell)

Pulse Interval Modulation

In the case of a pulsed signal, the length of the gap between pulses is also something that can be modulated, in the same way the pulse width can. Using the same analytic technique, the receiver would plot the number of pulse gaps versus gap duration and would see a finite number of clusters rather than a random distribution.

These two types of modulation can also be combined, so that one stream of bits is described using pulse width modulation, while another is described using pulse interval modulation.

Frequency Modulation

Frequency modulation is a straightforward and robust way to transmit information on a radio carrier. The frequency of the carrier is used to encode information, as in the simplified sound example at the beginning of this chapter.

Frequency-modulated signals are easiest to visualize in a frequency domain format, similar to a sonogram, as shown in the Fig. 6.9 below.

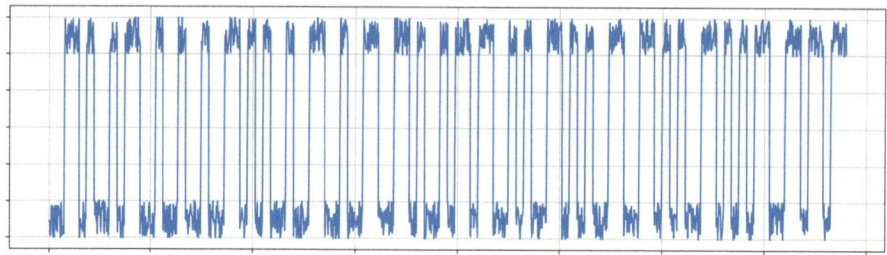

Fig. 6.7 A plot of a combined pulse width-/pulse interval-modulated signal with some background noise added in. See if you can work out what the transmitted message is. (Image credit: Brian McConnell. Data sets and answers can be found at github.com/aliencommunicationhandbook/exercises)

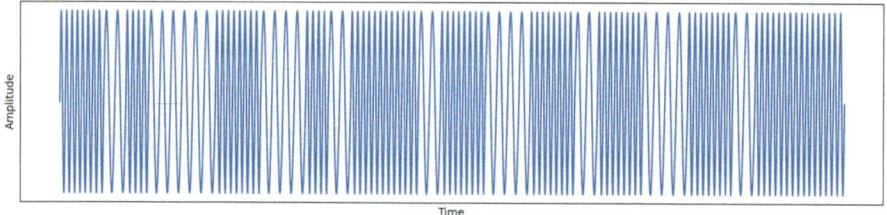

Fig. 6.8 A graph of a frequency-modulated signal. The x-axis represents time, while the y-axis represents amplitude. In this example, the amplitude remains constant, while the carrier frequency varies. (Image credit: Brian McConnell)

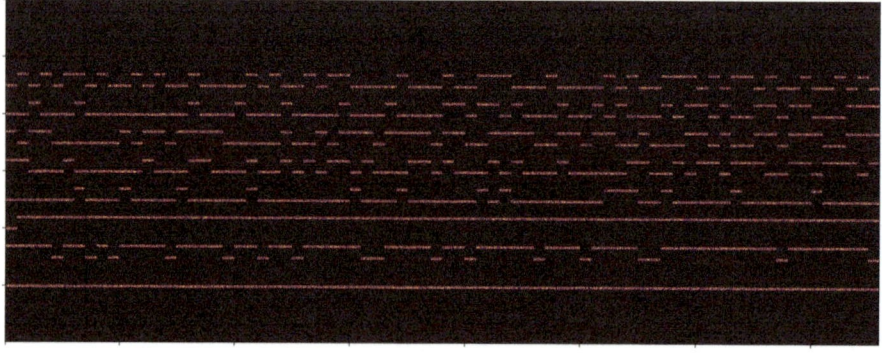

Fig. 6.9 A frequency domain plot of an FM signal with 8 modulated carriers, plus background noise added in. The horizontal axis represents time, while the vertical axis represents frequency. See if you can work out the content of this signal. (Image credit: Brian McConnell. Data sets and answers can be found at github.com/aliencommunicationhandbook/exercises)

Phase Modulation

This concept is trickier to explain to a layperson, so if you did not study electrical engineering, don't worry if you get stuck on this concept. Phase modulation works by varying the phase angle of the signal while holding the carrier frequency constant. The receiver can then combine this signal with an unmodulated reference signal to extract the data, which is revealed by the way the two signals interfere with each other. This system, known as *Phase Shift Keying*, is widely used in digital communication systems.

In addition to modulating the amplitude and frequency of a signal, it is also possible to modulate the phase of the signal. A monochromatic radio carrier is a sine wave, which can be expressed by the equation:

$$p = A \times \mathrm{Sin}\left(\left(f \times t\right) + \theta\right)$$

where A = amplitude.

f = frequency.
t = time.
θ = phase angle.

Phase-modulated information can be recovered by multiplying the modulated carrier with an unmodulated reference carrier that is matched to the same frequency. This multiplied signal is then averaged over time to reveal the encoded bits as a binary data stream, as shown in the Figs. 6.10, 6.11, 6.12, and 6.13 below.

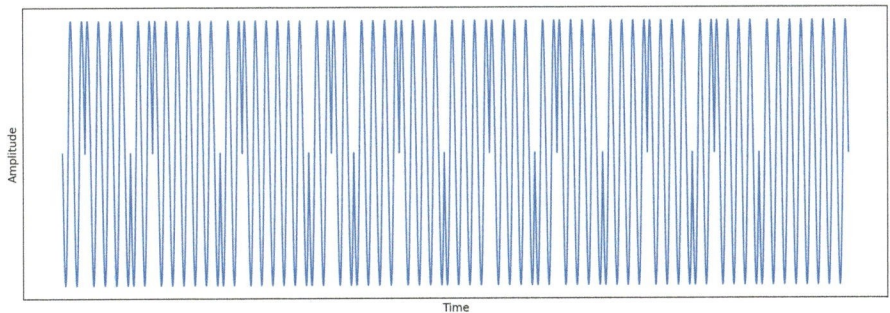

Fig. 6.10 A plot of a phase-modulated carrier. The x-axis represents time, while the y-axis represents amplitude. (Image credit: Brian McConnell)

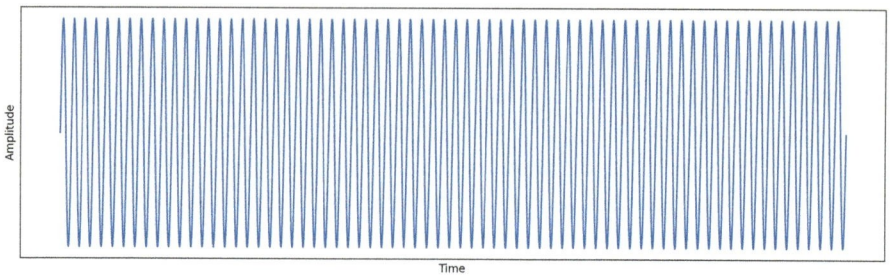

Fig. 6.11 A plot of an unmodulated reference carrier. The x-axis represents time while the y-axis represents signal power. (Image credit: Brian McConnell)

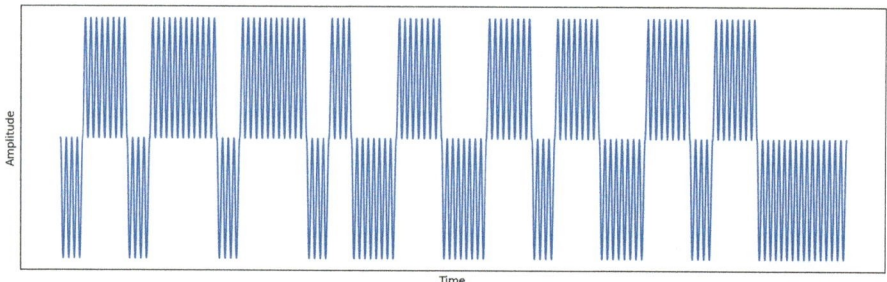

Fig. 6.12 A plot of the multiplied signals. When the carrier and reference signal are multiplied, the encoded bits are revealed. This signal can in turn be integrated (summed over time) to produce a digital signal or bitstream. (Image credit: Brian McConnell)

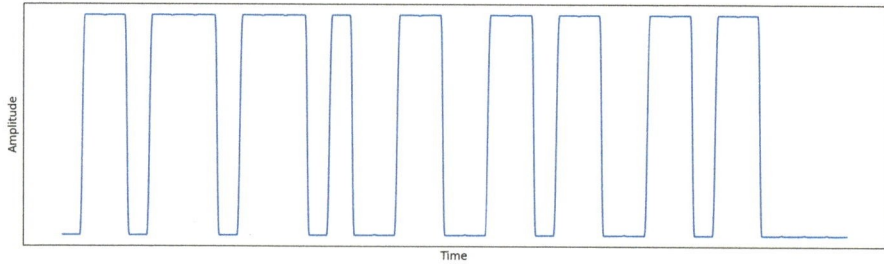

Fig. 6.13 A plot of the multiplied signal in Fig. 6.12 after being integrated over time. Notice that this signal alternates between two distinct states, a binary pattern. (Image credit: Brian McConnell)

Optical Signals

We've known since the 1960s Townes (1961) that lasers can be used to transmit information across interstellar distances. Lasers are an attractive way to transmit information because they are both *monochromatic* (they concentrate

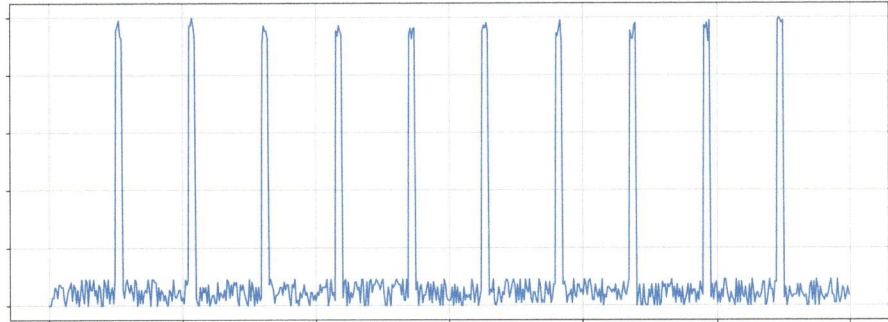

Fig. 6.14 A plot of the number of photons received per time interval. The x-axis represents time, while the y-axis represents the number of photons counted per time interval. A laser beacon can shine very brightly for a short time to outshine the background star during that brief time. Photons from the background star arrive at random intervals, while photons from the laser arrive simultaneously in a large group. (Image credit: Brian McConnell)

all of their power into a single wavelength or color of light) and because they can transmit extremely short pulses of light (less than a billionth of a second). This means they can outshine an entire star at a specific color or for a very short period of time.

The simplest type of OSETI setup involves the use of a pulsed laser to generate an extremely bright but extremely short pulse of light. This is akin to the flashbulb on a camera, which is timed to shine very brightly but only when the shutter is open, so the overall power requirement is minimized. An OSETI detector counts individual photons or parcels of light as they arrive. The photons from the background star will be evenly spread out over time, while the photons emitted by the ET laser will arrive at nearly the same instant. The result is a pulse in the photon count, which is clearly distinguished from the background.

This type of signal can be detected by ordinary optical telescopes if they are outfitted with an OSETI pulse detector. These detectors are typically configured with a beam splitter and two or more photon counters. The use of several counters allows the system to automatically ignore false positives generated by radioactive decay within the detector, cosmic rays, and other types of interference. Special-purpose telescopes, such as the PANOSETI instrument, that can watch large areas of the sky at one time are also under construction and will improve the sky coverage of optical SETI surveys.

So how might such a signal be modulated? There are several ways to do this, which can be used independently or combined to provide several independent communication channels on a single carrier. These include pulse length

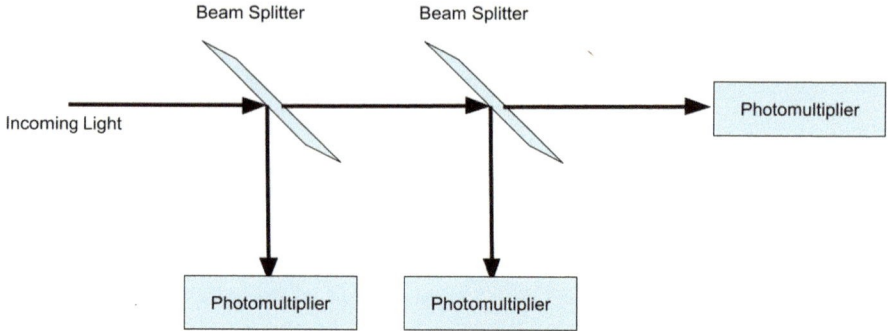

Fig. 6.15 A simplified schematic of a beam-splitting OSETI detector. The detector uses one or more beam splitters to split incoming light beam from the telescope out to multiple photomultipliers or photon counters. This design prevents false positives that occur in single detector systems (e.g., due to radioactive decay with the detector itself). Only if a blip is registered simultaneously in all detectors is it registered as an event. (Image credit: Brian McConnell)

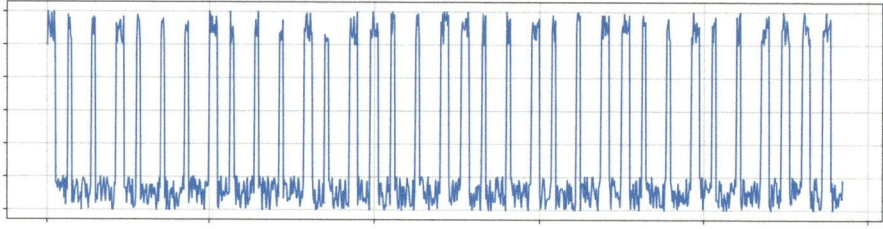

Fig. 6.16 A pulse width-modulated signal varies the length of a light pulse to encode information. In this graph, the slightly wider pulses represent 1 s, while the narrow pulses represent 0 s. (Image credit: Brian McConnell)

modulation, pulse interval modulation, wavelength modulation, and polarization modulation.

Pulse Length Modulation

The amount of time the pulse of light lasts can vary, so by adjusting the length of each pulse, the transmitter can encode a symbol. The simplest example of this would be a binary system where a short pulse represents a 0 and a longer pulse represents a 1; hence, each flash of light encodes one bit of information.

Suppose the sender wants to encode several bits of information in each pulse of light, using the pulse duration to do so. This can be done by allowing more than two pulse lengths. If there are four possible pulse lengths, the transmitter can encode 2 bits per pulse. Eight pulse lengths encode 3 bits per pulse and so on.

Signal analysts will plot the number of pulses counted versus their duration. If we see that pulse counts are clustered around certain durations, this will be a hint about the number of states that can be represented per pulse using pulse length modulation.

This will probably also hint at the type of numeric system used, since this is a low-level communication system. So, for example, if there are 2, 4, 8, or 2^n pulse lengths encountered, that's a hint that the low-level encoding scheme is binary, while if there is a different number of states, it might be using a different base numbering system.

Pulse Interval Modulation

In addition to varying the duration of each pulse, the transmitter can also vary the time interval between each pulse. This is particularly easy to do with a pulsed signal because the light source is off for the vast majority of the time, typically at least a million to one ratio between an on and off duty cycle. Just as with the pulse duration, the transmitter can describe the state using a finite number of intervals between pulses. A one-millisecond interval might represent a 0, while a two-millisecond interval represents a 1. And as with pulse length modulation, the transmitter can encode more than one bit of information per pulse by allowing more than two pulse intervals.

Once again, signal analysts can chart pulse counts versus pulse intervals to see if there are clusters around specific intervals and how many clusters there are. This will hint at the number of states that can be encoded with each pulse interval. Since a pulsed laser will be off something like 99.9999% of the time,

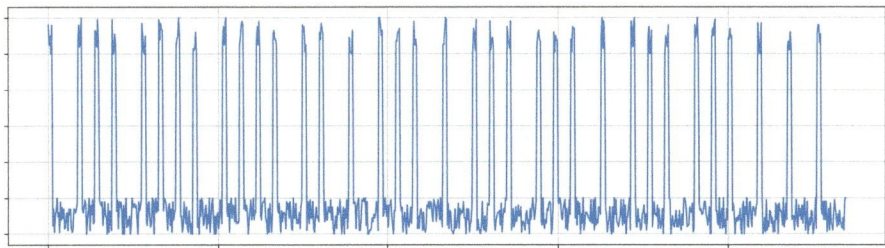

Fig. 6.17 Pulse interval modulation varies the time between pulses to encode information. In this graph, information is encoded in the gap between pulses, where wider gaps represent 1 s and narrow gaps represent 0 s. (Image credit: Brian McConnell)

and arrival times can be measured with considerable precision, it should be feasible to encode many bits of information per pulse interval.

Wavelength (Color) Multiplexing

Lasers are highly monochromatic, concentrating all of their power into a single color of light. The color of light is also something that can be modulated. A transmitter might be built with four lasers, each tuned to emit light at red, yellow, green, and blue colors and each of which can be fired independently. The four lasers are combined into a single beam. Thus, the color combination can encode several bits of information (in the four-laser example, this would be four bits of information, one bit for each color).

Another way of thinking about this is that each color of light is like another frequency on a radio. With this approach, it is possible to use color to create many separate communication channels, which can each employ pulse duration and pulse interval modulation to transmit information in parallel. The instrument required to receive this data would be relatively simple and would consist of a prism positioned in front of several photon counters, each positioned to capture different colors of light split out by the prism.

Polarization Modulation

Light can also be polarized. This refers to the direction that a photon oscillates in (a photon can be treated as both a wave and as a particle). The simplest form of polarization modulation alternates between horizontal and vertical polarization. In one modulation state, the photons emitted by the transmitter

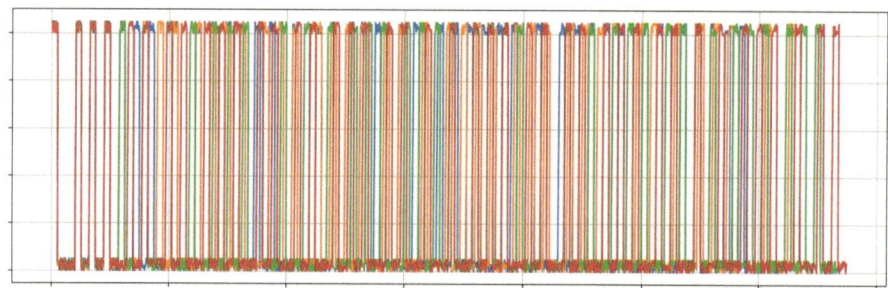

Fig. 6.18 A plot of four different color channels transmitting data using pulse interval modulation. By splitting the light out by color using a prism, one can separate these into independent communication channels. With a large number of color channels, it will be possible to transmit data at high data rates. (Image credit: Brian McConnell)

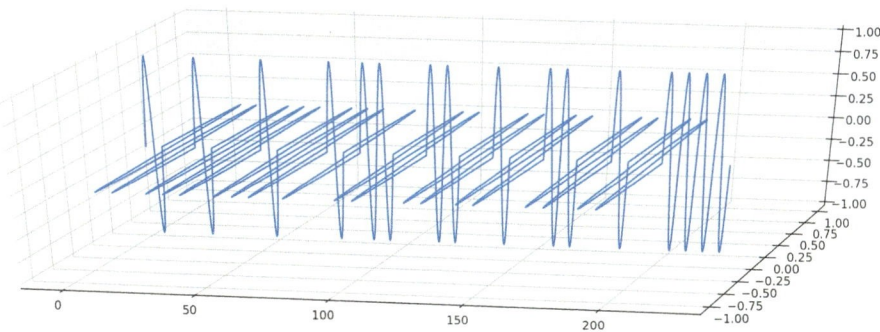

Fig. 6.19 A plot of a vertically/horizontally polarized signal. Notice how the sine waves alternate between up-down and front-back oscillation to encode a state. (Image credit: Brian McConnell)

oscillate up and down, while in the other state, they oscillate from side to side (or at a 90 degree angle relative to the other). Note that the definition of direction is arbitrary.

The receiver can detect the polarization of incoming light by splitting the beam to feed into a pair of pulse detectors, each with different polarization filters, similar to polarized sunglasses, attached to them. The transmitter can encode one additional bit in the polarization state. This form of modulation can also be thought of as a separate, parallel communication channel through which data can be sent. Radio signals can also be polarized, though it is simpler to illustrate how this will work for optical SETI instruments.

How Much Information Can ET Transmit?

The amount of information someone can send across an interstellar communication link is related to how much energy they use to transmit each bit of information, combined with the sensitivity of the receiving equipment. This involves a tradeoff between detectability and speed. One can either concentrate a lot of power in an easy-to-detect carrier but only be able to send information very slowly. On the other hand, one can send a lot of information by using less energy for each transmitted bit, at the cost of detectability on the receiving end.

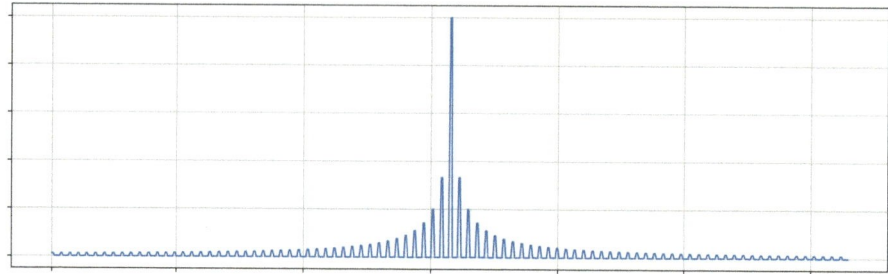

Fig. 6.20 A plot of a signal that consists of one strong central carrier, accompanied by many lower-power carriers. The horizontal axis represents signal frequency, while the vertical axis represents signal strength. The central carrier might be readily visible to our first-generation receivers, while it would take some time to build larger, more sensitive receivers that are capable of detecting and demodulating the lower power carriers. (Image credit: Brian McConnell)

Radio/Microwave Communication

In terms of radio or microwave signals, a good way to work around this is to design the system so that it combines a strong, easy-to-see signal with lower-power carriers that transmit most of the information at a higher combined rate. The receiver might not be able to see those lower-power carriers right away, but will suspect from the primary signal that there is something there. If we detect an ET signal, we will certainly examine the source carefully to look for other lower-power signals that went unnoticed during the initial detection.

The general strategy is to create a large collection of narrowband signals that are each modulated at a low rate so that the signal can be demodulated even if relatively long integration times are used by the receiver. If the sender has constraints on how much energy they can spend in total, they can transmit some carriers at high power, while using less power (less energy per bit transmitted) for side carriers. The graph above illustrates what this type of signal system might look like.

In aggregate, a system like this can transmit a lot of information yet do so in a way that trades detectability against energy efficiency. The receiver will have a good chance of detecting the powerful central carriers and demodulating the information carried by these channels. As they ramp up the scale and sensitivity of their receivers, they will be able to detect and demodulate the weaker side carriers and receive information at higher and higher rates. The total information carrying rate of the system can be estimated as the total number of carriers times their average modulation rate. Thus, a system of a million narrowband carriers, each modulated at one cycle per second, would deliver a million bits per second.

This type of design pattern also enables the transmitter to economize how much energy they spend. This can be implemented in many ways. Let's look at a simple pattern where we build up a system, starting with a single high-power carrier whose power budget is P. Then we add ten more carriers whose power budget is 0.1P apiece, so this second carrier group has a combined power budget of P. Next, we add 100 more carriers whose power budget is 0.01P each, for a combined power budget of P. We can continue this pattern of adding 10x the carriers with 1/tenth the energy budget per carrier. So for each 10x increase in information carrying capacity, the energy budget increases only by 1x.

Let's apply this pattern to create a system that transmits roughly a million bits per second. We would do this by creating carrier groups as follows:

1 + 10 + 100 + 1000 + 10,000 + 100,000 + 1000,000 for a total of 1,111,111 individual carriers.

The combined power budget would be 7P or just seven times the power budget of the strong, primary carrier. So instead of spending a million times as much energy, the transmitter is spending just seven times what it costs to operate the primary.

The tradeoff is that each successive group of carriers will be weaker and harder to detect compared to the primary. That's okay because this pattern is intended as an invitation for the receiver of the signal to keep looking for lower power signals. As the receiver succeeds in building more sensitive equipment, they gain access to new groups of carriers and can read out data at higher and higher rates.

Now, let's assume for the sake of argument that we find something like this, but due to limitations in our telescopes, we can only clearly see the primary carrier, and that it is modulated at about a bit per second. That is a glacially slow communication channel. What could anybody possibly communicate that would be interesting enough to grab and hold our attention? A bit per second works out to about 80,000 bits or 10,000 bytes of data per day. That's enough to send the equivalent of a 100 by 100 pixel image per day.

This is just one example, as there are plenty of other types of information one could send that would tip off the receiver that they are not sending random noise or numbers. Moreover, it is likely that we would see evidence of lower-power carrier groups, and while we might not be able to demodulate them right away, we would know where to look next and what type of equipment will be needed to make them readable.

Fig. 6.21 A downsampled (100 x 100 pixel) copy of the photo taken by the Apollo 17 astronauts, which has a raw file size of 10,000 bytes at 8 bits per pixel. In terms of information content, this is equivalent to what a 1-bit/second carrier could transmit in a day. Information like this would surely capture our attention. (Source image credit: NASA)

Interstellar Dispersion

One of the technical challenges in interstellar communication is a phenomenon known as *dispersion*. Light, when traveling through a perfect vacuum, will travel at the speed of c = 299,792,458 meters per second. But space is not perfectly empty. When light travels through matter, even a very diffuse gas, it will travel slower than c. The degree of slowdown is a function of the light's wavelength and how it interacts with the matter it encounters, such as ionized gas, interstellar dust grains, etc.

While this effect is small over short distances, it presents a problem for inter-stellar communication because some wavelengths of light will arrive at the destination noticeably sooner than others. For a broadband signal (one that is spread across a wide range of frequencies), the signal components will become smeared over time, which will make it difficult to recover the information encoded onto it.

This problem can be solved by organizing the signal as a very large number of narrowband carriers, each of which functions as a low-data rate channel that transmits information in parallel with the others. With this type of architecture, it does not matter if some channels deliver data between endpoints slightly faster than others. Another alternative is to de-disperse the signal on the receiving end, which can be done if the receiver knows the characteristics of the dispersion effect.

Dispersion itself can also be a sign of a technosignature, especially if the relationship between frequency and the amount of dispersion is the opposite of what would be expected from a naturally generated source.

This also hints at another pattern to look for: the parallel transmission of many data streams. Most discussions of an interstellar communication link tend to treat it as a serial transmission, which is analogous to a single pipe that all bits flow through, first in and first out. A transmission that is composed of many carriers, each transmitting at different rates or power levels, may not lend itself to the serial transmission of a single data stream. Instead, it will make more sense to transmit data in parallel.

This works especially well for data that is segmented into small parcels that can be stitched back together on the other end, in which case different segments can be transmitted in parallel via different carriers or carrier groups. One can think of this as many small pipes that carry information independently yet collectively transmit a large amount of information. In a system like this, it doesn't matter if the information sent through one channel arrives before information sent via another.

This signal architecture can also work around another problem, which is that some frequencies may be in use by the receiving site for other purposes (satellite communication, radar, etc.), obscured by atmospheric effects, etc. With this type of system, redundancy can be added so that segments of data are resent at different times via different frequencies.

We'll discuss redundancy and forward error correction in the next chapter.

Optical and Infrared Communication (Pulsed Beacons)

An interstellar optical communication system detects very short bursts of light that stand out against the light from the background star. This mode of communication lends itself best to pulse modulation by varying the duration of the pulse or the interval between pulses to indicate different signal states.

The transmission rate for this type of communication link can be calculated as follows:

$$R_{total} = N_{colorchannels} \times \left(r_{pulse} \times \left(b_{PDM} + b_{PIM} \right) \right)$$

Let's work an example to estimate the data rate for a relatively simple communication channel. The system employs 10 different color/wavelength carriers, each of which is modulated using pulse duration and pulse interval modulation. Each carrier pulses at a rate of 10 pulses per second. The pulse duration encodes 2 bits, for four possible pulse durations. The pulse interval has 1024 or 2^{10} values and thus encodes 10 bits per pulse interval. This yields a combined data rate of 1200 bits per second, comparable to an early dial-up modem.

The sender can increase the transmission capacity of the link by increasing the pulse rate or by increasing the number of color (wavelength) channels.

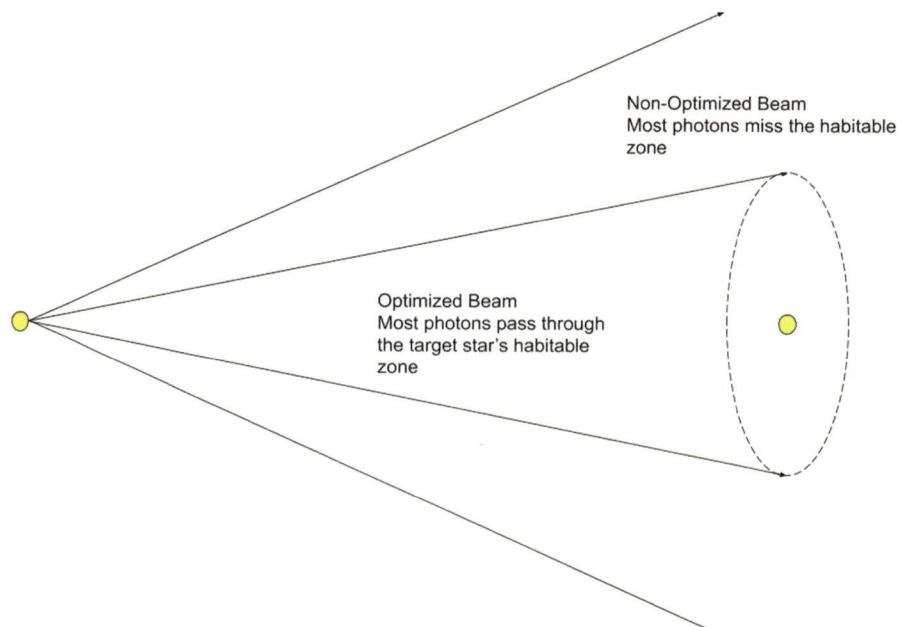

Fig. 6.22 An illustration of optimized and non-optimized beam. An optimized beam will be focused so most photons pass through the habitable zone of the target star, which reduces the amount of energy needed to produce a signal that can be detected at the destination

The primary constraint is how much energy the sender wishes to allocate to the transmitting system, as the energy required will scale in proportion with the pulse rate. We assume the system is already optimized to use the minimum amount of energy per pulse to allow detection at the destination, primarily by focusing its beam on the habitable zone around the target star. From a physics standpoint, transmission rates ranging from 10^8 to 10^{10} bits per second will be possible, although energy constraints and size requirements for transmitting facilities will likely result in lower practical data rates.

A pulsed optical signal can also be designed to maximize the chance of detection while also maximizing the information carrying capacity of the system. The sender could do this by using many color channels, each with different modulation rates, so that some channels are easily detected, while others are designed to transmit large amounts of data.

Let's consider a simple worked example where the sender transmits using red, green, and blue lasers. Each laser transmits data in parallel and can be thought of as a separate communication channel. Most of the time, they operate independently to maximize data transmission and use a high pulse rate with many bits encoded per pulse. Every few seconds, they go dark for a fraction of a second and emit a single bright pulse in all three colors.

While the multicolor signal may appear to be smeared together to someone whose detector does not split light about by color, they would be able to see the bright, combined pulses that appear every few seconds and would be able to read data out from these pulses with simpler detection equipment.

This strategy maximizes the chance that the receiver will notice the bright pulse train, even if their detector is not configured to split light out by color. They would be able to see the synchronized pulses with a monochromatic detector and to extract information from it, albeit at just a few bits per second. Not enough for in-depth communication, but enough to draw further attention to the source to look for additional modulation methods.

For more details, astronomer Seth Shostak of the SETI Institute provides an excellent discussion of the energy economics as they relate to transmission capacity in Douglas Vakoch's volume on *Communication with Extraterrestrial Intelligence* Shostak and Seth (2010).

Annual Transmission Capacity

If at first we are only able to detect carriers that transmit information slowly, we should not be discouraged. Even a slow connection, comparable to an early dial-up modem, is capable of transmitting a surprisingly large amount of information in the course of a year, as shown in Table 6.1.

Table 6.1 Channel speed versus the number of one megapixel images that can be sent per year

Bits/second	Bits/year	Photos/year (1 megapixel)
1	~32 megabits	4
100	~3200 megabits	400
10,000	~320 gigabits	40,000
1000,000	~32 terabits	4,000,000
100,000,000	~3200 terabits	400,000,000

Inscribed Matter and Artifacts

Another way an extraterrestrial civilization could deliver information to a nearby solar system is via inscribed matter, as we discussed in previous chapters and in particular Chap. 5 on Information Delivery (Carriers). This is akin to sending a CD-ROM via the mail. This method of information delivery is very slow, but it is possible to deliver extremely large amounts of information and also do so at a very low energy cost per bit of information delivered. As a complement to an electromagnetic signal, the sender could park probes within the target civilization's solar system. The probe's payload would be a durable material that is inscribed with information, similar to a CD-ROM but at much higher densities.

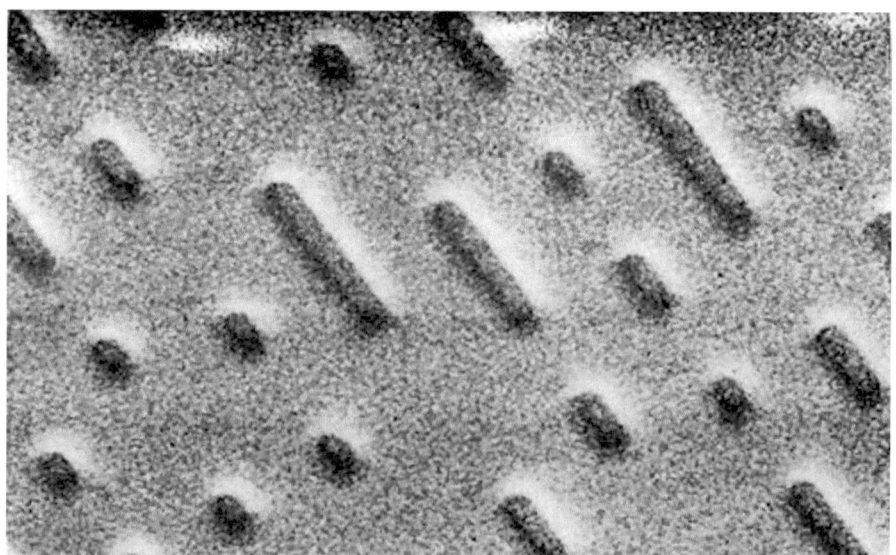

Electron microscope view of the pits etched into the surface of a compact disc, an example of inscribed matter. (Image courtesy: Zander Mausof). Mausolff, Zander. (2016).

Such a payload would be capable of delivering upward of 10^{22} bits of information per kilogram, a conservative estimate as DNA encodes information at a density of 10^{24} bits per kilogram. Information can be embedded in matter in a number of ways, such as etching microscopic pits into the surface of a thin layer of metal, as is done in CD-ROMs and DVDs. This can also be done by varying the composition of the material at the atomic level, for example, by varying the ratio of atoms used in a metallic alloy. Holograms are another example of a way that information can be encoded onto an object.

The amount of information that can be transmitted via electromagnetic radiation is limited both by energy constraints and by the spectrum that is suitable for long-range communication. If the sender is using microwave frequencies to communicate, they would be able to communicate efficiently using frequencies ranging from 1 to 10 GHz. Background noise becomes an issue above and below that range. This works out to a rough limit of about 10 gigabits per second in transmitting capacity if that spectrum were fully utilized.

That in turn works out to 10^{17} bits per year, five orders of magnitude less than the information that could be delivered with one kilogram of inscribed matter. The energy economics are also very attractive, with an energy cost that is far cheaper compared to radio- or laser-based transmission.

One can think of this as a cosmic version of "sneakernet." Before high-speed Internet connections were available, it was common practice for scientists to haul magnetic tapes and disk drives back and forth by foot or by car, hence the term sneakernet. While the probes might take thousands or tens of thousands of years to traverse interstellar space, their information carrying capacity, averaged over this time, would be far greater than an electromagnetic signal.

Who Wins the Race, the Tortoise or the Hare?

This may seem a bit counterintuitive at first, but inscribed matter, while it may travel very slowly compared to the speed of light, may be able to transmit much more information in a given amount of time. Let's compare a pair of hypothetical communication systems.

System A is an electromagnetic transmission channel that encodes one gigabit of information per second. This works out to 31.53 petabits (3.153×10^{16} bits) per year, assuming 24/7 receiver coverage year around and no redundancy for error correction (which is optimistic).

System B delivers information via inscribed matter and launches a one-kilogram probe that contains 10^{22} bits of information once every 100 years. This works out to 10 grams or 10^{20} bits per year on average.

The inscribed matter system, although it takes longer for information to travel to its destination, delivers information at an average rate 3000 times faster than the electromagnetic communication system. This probably understates how much information could be delivered via inscribed matter, for two reasons: (1) there is no upward limit on how many probes can be sent, other than the resource and energy budget of the sender, and (2) we could recover information from probes that were deposited in our vicinity eons ago, whereas information delivered via electromagnetic carriers is ephemeral and is lost once the signal passes by Earth. Because of this, the sender will probably need to send a repetitive transmission, which will undermine its information carrying capacity.

The disadvantage of inscribed matter probes is that they will be very hard to locate by chance. They would need to be compact and well shielded to protect inscribed matter from damage by debris and radiation. A one-kilogram probe would be small indeed, likely measuring a few centimeters on a side. It would look a lot like space debris, so unless the receiver knew exactly where to look for them, the likelihood of finding them by chance would be remote.

A more promising strategy would be to combine both delivery methods to leverage the strength of each. An electromagnetic signal could be detected by ground-based facilities without the need for space travel and could deliver significant amounts of information, though small relative to the amount of information an advanced civilization may have to share. The information in this signal can be used to point the receiver toward the location of inscribed matter probes and to transmit more time-sensitive information. This sort of combination strategy would combine a bright "we are here, look over there" signal with other modes of communication that are capable of transmitting large amounts of information.

Bracewell Probes

The sender could also employ Bracewell probes to combine the economics of inscribed matter with electromagnetic communication. A Bracewell probe, Bracewell (1960) like an inscribed matter probe, would be placed in the vicinity of the receiver, for example, in a high earth or lunar orbit that would remain stable for long periods of time, and when disturbed would begin actively transmitting data. It would then be able to transmit large amounts of information with a low power budget due to the greatly reduced distance between the transmitter and the receiver and could also interact with the receiver in real time.

The technology required to deliver a passive inscribed matter probe or a Bracewell probe would be very similar, except that a Bracewell probe's electronics would need to function for long periods of time, whereas an inscribed matter probe would not need to function after it had been placed in a stable orbit or deposited on a solid object such as a moon or asteroid. The best analogy would be to the Voyager space probes. They will remain functional for a few decades at most, but the gold records attached to them will be readable for over a billion years.

Conclusions

While it is possible that an information-bearing signal will appear to transmit information slowly at first, this could be the result of us missing other sub-channels or modulation schemes in the weeks and months following the initial detection. The limits on information carrying capacity are primarily a function of energy cost and can be mitigated by employing a multi-channel strategy that combines electromagnetic communication with the physical delivery of inscribed matter to maximize both detectability and the amount of information that can be delivered. That is to say, if an ET civilization wants to share a large amount of information, they will be able to do so without consuming extreme amounts of energy.

One thing that is clear is that inscribed matter will enable a civilization to deliver much larger amounts of information than radio or optical carriers can. This gap is made even greater by the fact that inscribed matter is a durable medium that, once delivered to its destination, will remain readable for millions of years or longer, whereas an electromagnetic signal must be captured at the instant it passes by its destination.

This is different from most narratives about contact, which typically involve some sort of interaction via signaling or physical contact (usually via an invasion). If inscribed matter is a prominent feature in how information is transmitted between civilizations, we may find that while the initial contact is made via a signal, most of the information is delivered via inscribed matter. The amounts of information that could be delivered this way would be so large that it might take years or decades just to read out and classify the information we have received, much less absorb and make sense of it.

References

Townes, C., Schwartz, R. **Interstellar and Interplanetary Communication by Optical Masers**. *Nature* 192, 348–349 (1961). https://doi.org/10.1038/192348c0.

Shostak, Seth. (2010). ***Limits on Interstellar Messages***. In Communication with Extraterrestrial Intelligence. Edited by Vakoch, Douglas A. SUNY Press. pp. 357–378.

Mausolff, Zander. (2016). **Experimental Determination of the Storage Capacity of a CD**. https://doi.org/10.13140/RG.2.1.4990.2968.

Bracewell, R. N. (1960). **Communications from Superior Galactic Communities**. Nature. **186** (4726): 670–671. Bibcode:1960Natur.186..670B. doi:https://doi.org/10.1038/186670a0. S2CID 4222557. Reprinted in A. G. Cameron, ed. (1963). *Interstellar Communication*. New York: W. A. Benjamin. pp. 243–248.

7

Lessons from Computing and Communications

What an extraterrestrial intelligence may choose to communicate is anyone's guess. However, the technical requirements for a reliable interstellar communication link are something we can understand and anticipate. Interstellar communication engineers will face a number of challenges, among them:

Disruption to line-of-sight communication – both radio and optical signals can be blocked for a variety of reasons. If the receiving civilization does not have telescopes positioned in multiple locations around its world, they will not be able to see the transmitting source for part of their day due to planetary rotation.

Transmission errors – if the signal is near the limits of the receiver's ability to detect it, they will not be able to read data from the signal with 100% accuracy. Some percentage of the message will be corrupted due to transcription errors. The same thing will apply to inscribed matter probes, some of which may be damaged or lost in transit.

Duty cycle mismatch – both the transmitter and receiver need to be pointed at each other at the right time. It is unlikely that a receiver will just happen to be looking as the transmitter starts broadcasting and will receive the entire transmission from start to finish. This is less of a problem for inscribed matter probes, but we need to find them first, and some may have been lost in transit.

Naive receiver – the receiver has not had prior contact with the sender and therefore has no prior knowledge about how the contents of the transmission are organized or what they represent.

B. S. McConnell, *The Alien Communication Handbook*, Astronomers' Universe, https://doi.org/10.1007/978-3-030-74845-6_7

With this in mind, we can understand the technical requirements in designing an interstellar communication link. A well-designed system should have the following properties:

- The system should allow the receiver to reconstruct a larger message from smaller parts received out of sequence.
- It should enable the receiver to detect and correct errors without requesting retransmission of incomplete or mis-transcribed segments.
- The basic structure of the transmitted data should be obvious, so the receiver can at least comprehend its basic structure, even if it takes time to understand the rest of what is being communicated.

In many respects, an interstellar communication link is just an extreme version of a wireless communication link, primarily due to the very long transmission delay caused by the speed of light and the likely gaps in reception. The latter makes it effectively impossible for the two parties to interact in anything approaching real time, for example, to request retransmission of missing or damaged data or clarification about something.

The first two requirements are key. It is important that the receiver be able to reconstruct the complete transmission from segments received out of sequence, because it is unlikely that they will just happen to be watching the right part of the sky at the moment the transmission begins or be able to receive it in its entirety in one pass.

It is similarly important that the receiver be able to detect and correct errors without requesting retransmission. This technique, known as *forward error correction*, is commonly employed in computers and communication networks. We will discuss it in greater depth shortly.

Segmenting Data

The solution for the first requirement is fairly straightforward, and that is to divide an arbitrarily large block of data into smaller parcels of data and label them each in such a way that they can be reassembled, even if they are received out of sequence. Think of this in terms of a book that you want to send to someone. You could send the book in one package, but risk losing the entire shipment in transit, or you could send the book in sections, for example, by mailing one chapter at a time. You could take this example even further by mailing one page or one paragraph per envelope and include instructions about how to reassemble the book on the receiving end.

When segmenting data, you take a block of data and then append extra data, known as *metadata*, that describes the payload being delivered in each parcel. This metadata can include things like:

A segment number – this allows each segment to be identified and where it is in a larger sequence of segments that can be used to build a collection. This is akin to a page in a book.

A collection number – this identifies a larger collection that the segment belongs to, akin to the title or ISBN number of a book.

A parent collection number – this identifies a collection or collections that the package belongs to, akin to a bookshelf with a collection of related books.

A sender and/or receiver ID – if there are multiple participants in a network, you may also want to identify the sender and receiver.

Error detection or error correction codes – the data may be encoded in a redundant manner to enable the receiver to detect and correct transcription errors without requiring information to be retransmitted (see Forward Error Correction).

In plain language, a system like this says: "I will now send N bits of data. This is segment X out of Y segments. This segment also belongs to collection number A, which belongs to collection numbers B, C and D." This extra information is all the receiver needs to stitch these smaller blocks of data into a larger collection that in turn belongs to an organized collection of collections. It is also worth noting that this allows an organized collection to be built in a bottom-up fashion, as the transmission is received in parts.

The cost of adding this metadata can be small relative to the amount of information being transmitted. We can show this with a worked example. Here, the sender segments each frame of data as follows:

Sender ID: a 96-bit address that uniquely identifies the sender
Receiver ID: a 96-bit address that uniquely identifies the receiver
Object ID: a 96-bit address for the object/collection ID (similar to a filename)
Parent ID: a 96-bit address for the parent collection the object belongs to (similar to a parent tree node or directory tree)
Segment ID: a 32-bit number representing the frame number or segment within an object (similar to a page in a book)
Type ID: a 16-bit address that identifies the media type of the object (e.g. mono color/grayscale image)
Frame Length: the number of payload bits that will follow the header/metadata (16 bits, can be 0–65,535; let's assume the sender transmits about 16 kilobits per frame)

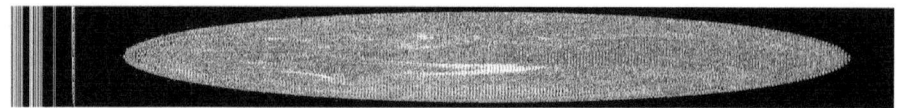

Fig. 7.1 An image with metadata prepended to each row including a file or collection ID, sender ID, data type ID, and segment or row ID. Notice the metadata to the left side of the bitfield. See if you can work out what this is. (Image credit: Brian McConnell. Data sets and solutions can be found at github.com/aliencommunicationhandbook/exercises)

The 96-bit address fields create a large address space. This allows 10^{29} unique objects to be referenced in the address space, an astronomically large number. The numbers chosen for the field lengths in this example are arbitrary and are probably larger than they need to be, but they show how it is possible to have a very large number of uniquely addressable items in a hierarchical collection. Yet the "tax" of adding this metadata is only about 2.5% in this example.

An added bonus feature of this system is that the metadata prepended or appended to each block of payload data has a repetitive structure that will be detectable via statistical or even visual analysis, as demonstrated in Fig. 7.1.

What you end up with is something that looks a lot like a computer network or file system. This isn't to say that ET will copy the systems we use on Earth, but it would not be surprising if the communication system has some of these characteristics, so segmentation is a pattern to be on the watch for.

Forward Error Correction

The sender will have to assume that the communication link is not reliable, meaning that some percentage of data is lost or corrupted in transit for a variety of reasons. In computers and digital communication systems, *forward error correction* enables errors to be detected and in many cases to be corrected without further involvement by the sender. This is an important requirement for interstellar communication because of the long communication delays due to the speed of light.

Let's go back to the analogy of mailing a book to someone one page at a time. Imagine there is a chance that each envelope will be lost in transit and that we want the recipient to be able to reconstruct the book if some envelopes are lost. An easy way to do this is to send each page more than once, so that if a page is lost in transit, the recipient will probably receive one of the other copies. This is a simple analogy of how forward error correction works.

One of the simplest error correction codes is the so-called majority vote code, also known as an N-modular redundancy code. Here, you use an odd number of bits to represent a single bit. In a (3,1) code, 000 represents a 0, while 111 represents a 1. If one digit in the triplet is flipped, the receiver can detect that an error has occurred and recover the lost data. The sender can further increase error resistance by adding more redundancy, for example, by using 5 or 7 bits to encode each bit of information.

The tradeoff with this approach, and with forward error correction in general, is that error resistance comes at the expense of information carrying capacity. Using a (3,1) code as shown above reduces the information carrying capacity to a third compared to unencoded data, since it uses 3 bits to represent each bit of payload data.

If a majority vote code is employed in a message, this will be easily noticed in statistical analysis. If a (3,1) code is utilized, what you'd expect to see is an unusually large population of triplets (000 or 111) in the transcribed data stream versus a much smaller population of other combinations (001, 010, 110, etc.).

The main weakness of the majority vote code is that while it is good at correcting widely dispersed single-bit errors, it is not able to deal with burst errors or prolonged dropouts in communication. So how can one design a workaround to that problem?

This is an example of where data segmentation solves several problems at once. Because the sender can transmit segments of data out of sequence, they can also send them more than once at different times. The more important a particular piece of data is, the more it can be resent. This is a way of creating redundancy over time, so that if the receiver misses an important segment of the message, they can grab the missing segments the next time they are transmitted.

This approach forms the basis for a forward error correction system that is both robust and whose method of operation is easy to understand. Let's suppose two identical frames of data are repeated three times over several days. This is a temporal version of the (3,1) majority vote code described earlier in this chapter. When the three frames are compared, the receiver would see a table like this.

This approach combines the simplicity of the (N,1) majority vote code with repetition over time to compensate for short- and long-duration disruptions to the transmission. This is also a form of error correction that can be added without making the message much harder to parse. The receiver will notice that some parcels of data are being resent several times. Besides providing a temporal form of error correction, this is also a good way to hint at

```
1111110000000000101000000000111111111111111111111111111111100111000000001100000
0111000111111111111110001110001111110000000000111111111011111111111111110001111111
1111100011111111111111111111111111111111111111000000111111111111100000011111 1
111111111000111111111111111111000000111111110001111111111111000111111100011110111
10001111110001111111111111100000011111100011111111111100000000000001111110001111
111111111111111111111111111111101011111100011111100000000001110000001111111 1
11111000111111000111111111100000011111100011111111100011111100000001011111100
11111110001110001111111110001110000000111000111111111111111111111111110000001
11111111000000111000111000111111000000111000111111000000011100000000011111000
1111111111110001111111111111111000000000011111100011100000000001111111111111100
011111111100011100011111111100000011111111111111100011111111111111111111111 1
11000111111111111111111100011111111100011111111100000001010001111111111000111111
111111000111111000000111000111111111111111000111111111111000111111111111100011
111100000001011110000000111111111111111000000000011111111111111111111111111111
1111100000000000000111111000111111111111111000111000111000111111000111111111000
111111000111111111100011100011100011111100011111111011000000101100011111100011
11111110000000111111000111000111111111100011111111111111111111111111111110000000
0000011111111100011111111100011111111111111111111111100000001110001111110111111111
0001111101111111111110001111111111110001111111111111111111111111111100011110100
01110000000111000111000111111000111000111000000011101111111111111111100011111110
00000111111111000000111111111111100011100011111100011100011100011111100011 1000
111000000011111111111111111111111000111000111111111111000000011100011100011111100
0000111111111100011111111100011111111111111110000000000111000111111111111111111111
1100011111100011100000001110001111111111000000011111111111111111000111000111111111
111000000000111111111111111111111111000111111111100011111111111111111111111100011
1000111111111111111111100000000000111111111100000011111111111111111111111111111110
0011111100011111111111111111111110001111111111111111100111111100011111111111111111
1111110001110000000001111111111111111100000000111111111111000111000000011111100000
0111111111111111111111111111111000000111111111111110001111111111111111111111111110001
111110001111111111000000111111111111110001111110001111111111111110001110001111111111
11111100011111111111111111111111111000111111111100011100011111111110001111111111
10000001111111111111111111111111111111111111111111111110001111111111111111111111111
1100011100000000011111111111111111111000111000111111000111111111111111111000111000
00011100011111100000001111110001110001110000000111000111111111111111111111111111100
0111000111000111111111111111111111111111111110000111111111111111111111111111111111111
11111000000011111100011100011111111111111111111111111111100011111111111101110000000
0001110001111111110000000111111111100011111100011111100011111111111111111111100011
100011111100000011100011111111111111000110111000111000111000111000000011111111
1111100011111111110001111111111111110000000111111111111111111111111111111111111111000
0001110001011111111000111111111100000000000000000001111110001111111000111111
```

Fig. 7.2 An example of a bitstream with triple modular redundancy applied to it, along with a simulated bit error rate of 1%. Data sets and solutions can be found at github.com/aliencommunicationhandbook/exercises

Table 7.1 An example of (3,1) error correction applied to segments that are retransmitted at different times

Segment 1217, Sent on Day 1	0	1	1	0	0	1	1	0
Segment 1217, Sent on Day 2	0	1	0	0	0	0	1	0
Segment 1217, Sent on Day 3	0	1	1	1	0	0	0	1
Segment 1217, Reconstructed	**0**	**1**	**1**	**0**	**0**	**0**	**1**	**0**

which specific parcels of data are most important (more repeats → more important).

Temporal error correction is important because the receiver may not have continuous sky coverage. Imagine a mostly oceanic world that has one continent covering a small part of its surface area. Their land-based telescopes would only be able to see the transmitter for a fraction of their day. Temporal error correction will enable a receiver with only partial sky coverage to reconstruct the full data stream.

There are much more sophisticated ways to encode data for forward error correction, but these generally involve computer programs to implement, which we will discuss in Chap. 9 on Algorithmic Communication Systems. The ability to implement robust and sophisticated error correction protocols is one reason to expect that the transmission will include computer programs.

How Much Error Correction Is Enough?

Error correction is especially important when the sender needs to protect against single-bit errors. When you run a computer program, a single-bit error in the underlying program code can cause it to fail catastrophically or crash. Error correction is a good thing, and it is necessary for some media types.

There are many situations where errors are less problematic and can be tolerated, provided there are not too many of them. Consider a digital photograph. The simplest way to represent a photo in a digital format is using a two-dimensional array of numbers. This is known as a *bitmap*, and each element of the array represents a pixel or picture element. The relative brightness of each pixel is defined by a number that ranges from 0 to a finite value.

In a situation where each bit in the data set has a small chance of being corrupted (flipped from a 0 to a 1 or vice versa), this introduces a small amount of noise where random pixels appear to be brighter or dimmer than they should be. Most of these random errors will affect less significant bits, so the effect of those errors will be small. About half will affect more significant bits, which will cause those pixels to be unusually bright or dark relative to the correct value. This gives a photo a speckled appearance. Provided the error rate is low, the effect is tolerable.

The point here is that some types of media will require error correction, but others, such as observational data representing things like photos or sound, may not, especially if the average error rate is low. Error correction may be applied to some regions of the message and not others in order to maximize the amount of information that can be sent.

Fig. 7.3 The images above show the photograph of Earth taken by the Apollo 17 astronauts. The leftmost image has no bit errors. The middle image has a 1% bit error rate. The rightmost image has a 5% bit error rate. Notice how the images with errors have a speckled appearance but are otherwise intact. (Source image credit: NASA)

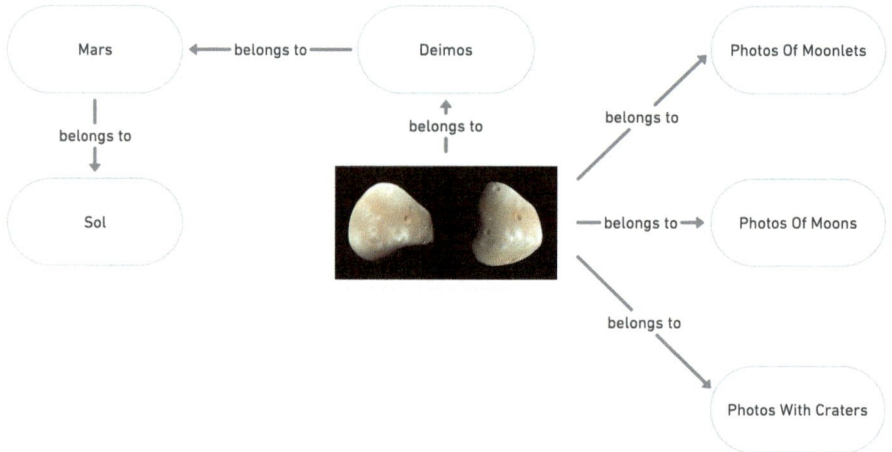

Fig. 7.4 A diagram that describes which collections an image belongs to. This sort of bottom-up notation can be used to build up a collection of collections as the transmission is received. (Image credit: Brian McConnell)

Building a Tree

One way to organize information into a collection of collections is to build what's known as a tree. To do this, each member of a collection describes which parent collection or collections it belongs to. This allows members of a collection to be described in no particular order and allows the recipient to build up the overall tree structure as they receive the transmission.

This is similar to the way a computer file system works, so we can describe this in English notation for readability. Let's say we want to include a picture of the Martian moon Deimos and indicate which collections this image belongs to.

By labeling data parcels so they describe which collection each parcel belongs to, and which collection that collection belongs to, the sender can provide all of the information needed to build up the equivalent of a directory tree as the transmission is received. This differs from how a computer file system is typically defined, as that is done in a top-down manner, where one creates directories and subdirectories and then adds files to them. With this approach, a collection can be mapped to many other collections, so it can be defined once but appear in many collections.

Hinting at Message Structure

There are a number of things the sender can do to hint at higher level structure within a data stream and can do so without bloating the message a great deal. Structure is important because it allows the sender to define boundaries between collections and to describe more complex data sets.

Padding is one way to do this. This involves adding extra digits to hint at boundaries between subelements or fields in a larger parcel of data. Let's say that the sender appends two numbers to a larger block of payload data. One number is a collection ID, which as we discussed is a unique number that refers to a collection of message segments (analogous to a file on a computer or the ISBN number of a book). The second number is a segment ID, which as discussed is a counter that increments from 1 and identifies the sequence of an individual segment, kind of like a page number in a book.

Normally a network designer will want to minimize the number of bits allocated for metadata fields such as those described above. This is known as *an address space*. The address space only needs to be large enough to avoid address conflicts. Until recently, the Internet used a 32-bit address space, which is enough to allow for about 4 billion unique addresses. This was more than enough in the 1960s, but it is not enough in today's hyperconnected world, so this is being increased to a 128-bit address space. The table below shows how the size of an address field relates to the number of unique addresses it can represent.

An analyst looking at the data will see that for every N digit, there is a long repeating series of mostly zeroes. While it won't be immediately obvious what this represents, the periodicity of the pattern will be obvious and will be a way

Table 7.2 Address size versus the number of uniquely addressable items

Address size in bits	Unique addresses (2^{bits})	Remarks
8	256	1 byte
16	65,536	2 bytes
32	4,294,967,296	IPv4 address space, 32-bit addresses or int/float values on typical computers
64	1.8×10^{19}	For comparison there are an estimated 10^{21} stars in the observable universe
96	7.9×10^{28}	The number of molecules in one cubic meter of water
128	3.4×10^{38}	IPv6 address space
256	1.1×10^{77}	For comparison, there are an estimated 10^{78} to 10^{80} atoms in the observable universe

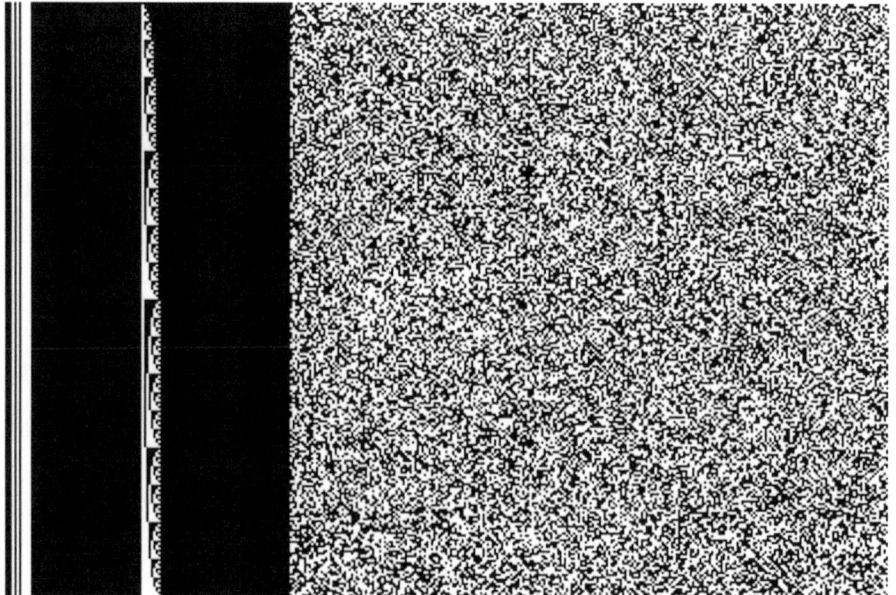

Fig. 7.5 A simulated bitstream padded with metadata where one metadata field contains a collection ID, and the next contains a segment ID, followed by 512 bits of efficiently encoded data. Notice how the collection ID (to the far left) remains constant, while the next field increments by 1 for each row. (Image credit: Brian McConnell)

to hint that the data is organized into an array. This comes at the expense of increasing the amount of data that needs to be sent but at most by a few percent.

Another way to do this will be to introduce gaps into the transmission itself, so the transmitter appears to cycle on and off at regular intervals. It might transmit a few thousand bits of data, go silent for a fraction of a second, and then start up again. This is another way to hint at the boundaries between collections of digits within a larger data stream and is similar to padding. It reduces the information carrying capacity of the signal somewhat but also makes it easier to see structure in the transmission. The length of the pause can be used to hint at nested structures, where a short pause might indicate a boundary between fields within a row of data, while a longer pause might indicate a boundary between rows. Fun fact: this is how analog televisions kept horizontal and vertical scan lines for pictures in alignment.

Structure can also be described within the modulation scheme for the signal itself. Let's imagine that a carrier signal chirps at one of the four frequencies, meaning that each chirp maps to one of the four symbols, A, B, C, and D, two of which map to the numbers 0 and 1 and two of which map to the equivalent of open and close parentheses (and). This scheme would enable the sender to transmit richly structured data, such as an n-dimensional array of numbers.

Example: a two-dimensional array of numbers

```
( ( (1) (10) (11) ) ( (100) (101) (110) ) ( (111) (1000) (1001) ) )
```

Translates to

```
1    2    3
4    5    6
7    8    9
```

The point isn't to predict which specific method a sender will use to do this, just that we should be looking for patterns like this.

Checklist of Patterns to Look for

Is the signal organized around a single carrier or around many carriers that are modulated to transmit information in parallel?

Are the carriers all modulated at the same rate, or are some modulated at different rates to encode information at different rates for some carriers (e.g., to combine a high-power carrier that transmits information very slowly with side carriers that transmit information at higher speeds)?

Are many modulation schemes used in parallel (e.g., amplitude, frequency, polarization, etc.)?

Are there repetitive or sequential patterns within the data stream (e.g., sequences that translate to numbers that are being incremented by a constant amount)? Are there long sequences of identical states (this is a pattern we'd expect if padding is in use)?

Does a carrier appear to go quiet on a regular basis? This could be used to hint at boundaries in structured information.

Are all decoded signal states encountered with the same probability, or is there a bias in their usage? For example, in the case of a carrier that encodes four possible signal states, are some states used more often than others? This might hint at the use of some signal states to communicate data structure, while others communicate data or operands.

8

Entropy: Measuring Order and Randomness

B. S. McConnell, *The Alien Communication Handbook*, Astronomers' Universe,
https://doi.org/10.1007/978-3-030-74845-6_8

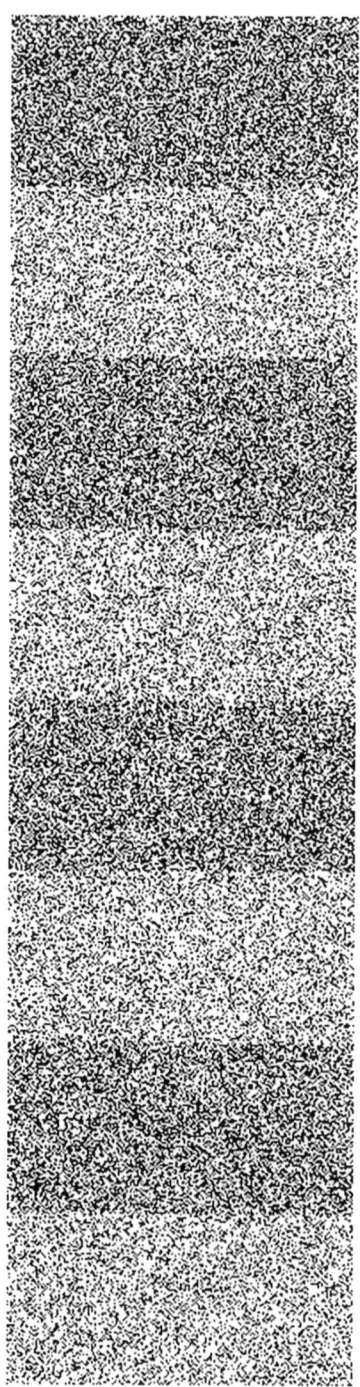

One of the first things analysts will begin to do as data is transcribed from an ET signal or artifact is to examine its *information entropy* or *Shannon entropy*. This is a measure of how many bits of information are represented by each symbol or state within a message or information density. Another way to think of Shannon entropy is as a measure of "surprise-ness".

The areas of the message that contain lots of repetition, white space, and obvious structure will have a low information density, and low entropy, relative to parts of the message that are efficiently encoded. Efficiently encoded regions will look more like random noise.

Low-density regions are interesting because this is where we are most likely to find more easily decoded information, such as uncompressed images, diagrams, etc. This form of analysis will work for any series of digits or symbols and can be used to map the information density of any digital signal, whether its code is binary or any other numeric base system.

The banner in this chapter displays a simulated bitstream that alternates between two regions. In one region, each bit has a 50/50 probability of being a 0 (black) or 1 (white). This is a random distribution and is what efficiently encoded data will look like. In the other region, each bit has a 70% probability of being a 0 (black) and a 30% probability of being a 1 (white), meaning there are more zeros than ones. It is easy to see the difference between the regions as one appears darker than the other.

When we map the entropy of this bitstream, we will see that the random regions have the highest information density, with 1 bit per pixel, while the non-random regions are less information dense. The non-random regions will be of particular interest because they are more likely to contain repetition and more obvious structure that may be of use in understanding the more efficiently coded parts of the message.

Even without knowing what the encoded information represents, we can map its information density and use this to characterize different regions of the message by how much information they contain.

Figure 8.1 charts the first-order Shannon entropy of an English text that has been padded on either end with random letters. This is the type of pattern we would expect to see when there is a region of more ordered content that is surrounded by more efficiently encoded information, such as compressed data. Even without knowing anything about what the information represents, it is possible to use an analysis like this to spot the regions that have higher or lower information density.

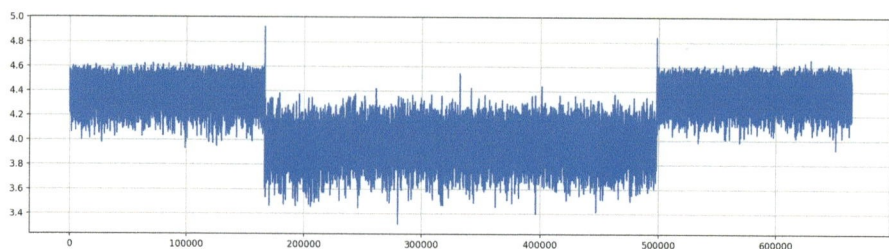

Fig. 8.1 A graph of the first-order Shannon entropy, H_1, of an English text. The beginning and end of the text is padded with random letters, while the middle region consists of text from Mary Shelley's Frankenstein. The x-axis represents time, while the y-axis plots the number of bits of information per symbol. (Image credit: Brian McConnell)

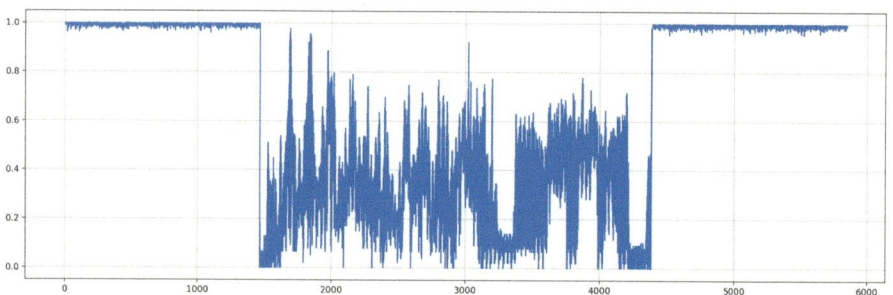

Fig. 8.2 A plot of the first-order entropy of a bitstream that contains a black/white bitmap, surrounded before and after by random ones and zeroes. This is similar to what we would expect to see if an uncompressed image is interleaved with more efficiently encoded or compressed data. Notice that the efficiently encoded regions have higher information density than the region in the center of the plot. (Image credit: Brian McConnell)

Next let's look at the entropy of a bitstream that includes a line drawing surrounded by what appears to be random data. This is the type of pattern we might expect to see if uncompressed images are interleaved with a collection of images that have been compressed to minimize the amount of data needed to represent them.

Next, let's repeat this example by measuring the entropy for a bitstream that contains a grayscale image, also surrounded before and after by random digits.

Here is another example. This time, a digitized audio sample is embedded in a larger data stream.

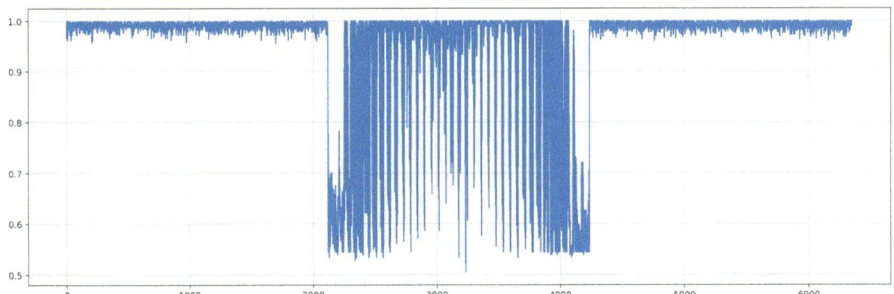

Fig. 8.3 A plot of the first-order entropy of a bitstream that includes an uncompressed grayscale bitmap of unspecified encoding. It is surrounded on either side by random ones and zeroes, which is what we would expect if an image like this were interleaved with more efficiently encoded data. (Image credit: Brian McConnell)

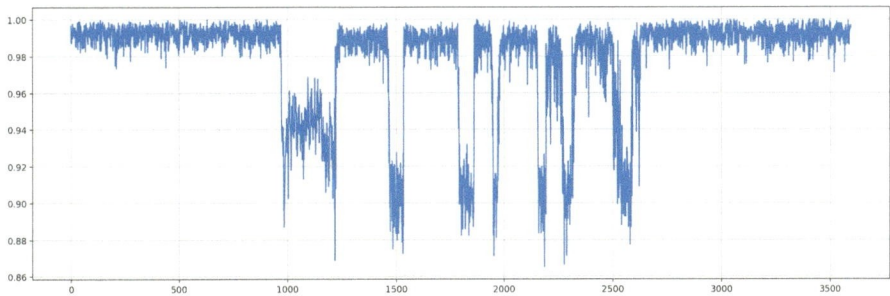

Fig. 8.4 A plot for the first-order entropy of a bitstream that includes a digitized audio sample of unspecified encoding. It is preceded and followed by more random data. Notice how the H value, or information density, changes over time. (Image credit: Brian McConnell)

Next let's take a look at what a segmented data stream might look like. Here we simulate a data stream where the data is segmented into N bits of payload data. Each chunk has a 64-bit header attached to it that is akin to a page number. The payload bits are nearly random, to simulate efficiently encoded data.

These examples show that we can learn about the structure and content of a data stream by studying how its Shannon entropy changes over time. More orderly segments of the data stream will have a lower H value or bits encoded per symbol compared to regions that appear to be more random.

What is interesting is that we ran unstructured bit streams through this analysis. No assumptions were made about byte length, the types of encodings used for images, audio, etc. While this sort of analysis won't tell us anything about the higher-order structure of the data or its meaning, it will give us an

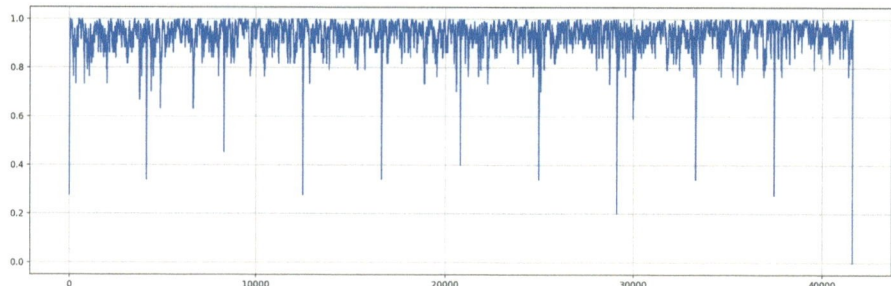

Fig. 8.5 A plot of the first-order entropy for a data stream that is segmented into smaller parcels, each of which has a 64-bit header prepended to it that is less random than the payload bits. Notice the repetitive pattern where the header regions appear, as they have a lower H_1 value than the payload segments. (Image credit: Brian McConnell)

idea of where to look for information that is easier to parse, such as uncompressed images, audio, or primers.

9

Algorithmic Communication Systems

© The Author(s), under exclusive license to Springer Nature Switzerland AG 2021
B. S. McConnell, *The Alien Communication Handbook*, Astronomers' Universe,
https://doi.org/10.1007/978-3-030-74845-6_9

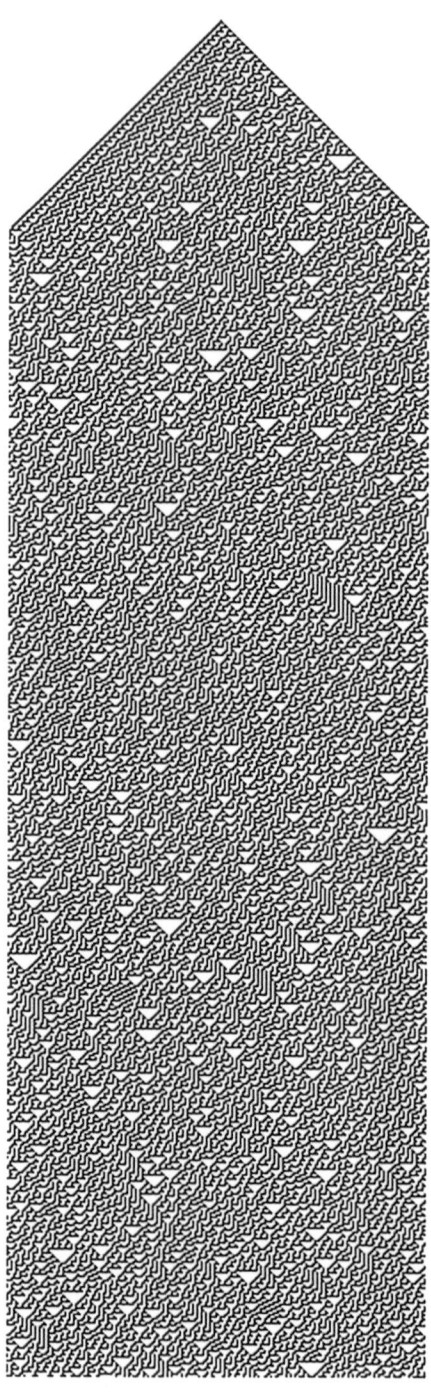

Fictional treatments of alien contact scenarios often envision the message as a static entity, not unlike the Egyptian hieroglyphs. We should seriously consider that a message may be composed of computer programs. There is an important reason for this, and that is to work around the time delays and other constraints associated with interstellar communication.

An interstellar communication link is likely to be unreliable and slow compared to short-range communication systems. The speed of light also ensures that a conversation, even with our neighbors at Proxima Centauri, would involve multi-year delays – 8 years at a minimum. Knowledge about digital computing is a de facto requirement for success in establishing a communications link, so it's not unreasonable to assume that the sender will know about computing and may incorporate algorithms into a transmission.

Algorithms, the procedures implemented in computer programs, are useful because they can implement arbitrarily complex instructions and math operations using a small and simple set of basic math and logic symbols. A program could represent something as simple as a tic-tac-toe game or something as complex as a climate model or machine learning system. The limits of communication are only constrained by the imagination of the sender.

The data banner in this chapter displays the output of a rule 30 cellular automaton, a simple algorithm that produces results that appear both orderly and chaotic.

An algorithmic communication system is attractive because the amount of information that can be transmitted across an interstellar link is finite and possibly quite small relative to the amount of information the sender might want to share. If the sender can compress data using an algorithm and includes an algorithm that can be used to decompress it on the receiving end, the sender can increase the effective carrying capacity of the link manyfold.

A wise strategy will be to send parts of the transmission as plain text, without compression, and parts that are intended to be processed by a decompression algorithm on the receiving end. That way if the receiver is unable to comprehend the algorithmic part of the message, they have a decent chance at being able to parse the static, uncompressed parts. If this pattern is in use, the receiver should notice that some regions of the data stream have higher information density than others.

It is also possible that programs could be used to teach concepts that are difficult to portray with a static image or mathematical formula. Turbulent flow in a fluid is an example of this; it is difficult to explain with a still photograph or mathematical equation. A motion picture image can better describe turbulence but requires a large amount of data, whereas a compact program can simulate turbulent flow in a wide range of conditions. For example, this web-based turbulent flow simulation by Dean Alex (2013) equires just 11 kilobytes of JavaScript code to implement. See Chap. 15 for more on turbulent flow.

ET Basic

What are the minimum requirements for an algorithmic message? An algorithmic message is essentially a computer program. Computer programming languages, like human languages, boil down to a small vocabulary of fundamental symbols. Even the largest and most complex programs are composed of these basic instructions, which include operations such as:

- Boolean logic (AND, OR, NOT)
- Math operations (addition, subtraction, division, etc.)
- Comparison (is condition A true or not; is number A larger than number B?)
- Branching (if condition A is true, do this, otherwise do something else)
- Read or write data from memory, for future reference, or to share data with other programs or an external interface
- Abstraction (define a reusable segment of code and reference it via shorthand elsewhere)

While the basic vocabulary of operations is small, these can be combined in an infinite variety of ways. Every computer program you have used, whether it is a simple tic-tac-toe game or a smartphone app that can transcribe as you speak, reduces to a simple instruction set like this.

There are a few basic paths to defining the foundation of an algorithmic message. One approach is to describe instructions that will run on a virtual or imaginary computer. This is known as an *interpreted language*, where the language is unrelated to the computer hardware it runs on. This type of programming language can be described by sending examples of simple instructions, where the receiver is prompted to guess the meaning of an unknown symbol within an otherwise known example. How the receiver chooses to implement things in hardware is left up to them.

Let's say we want to teach the recipient a symbol for addition. The recipient has already learned how to read numbers in a previous lesson, so we introduce a new symbol ⇑ to represent addition. In a machine-readable code, this symbol would be a unique numeric identifier, but we'll use more familiar notation here for readability. The example sequence would contain a series of statements like:

$$1 ⇑ 1 = 2$$
$$1 ⇑ 2 = 3$$
$$2 ⇑ 1 = 3$$
$$2 ⇑ 2 = 4$$
$$4 ⇑ 5 = 9$$
$$5 ⇑ 4 = 9$$

With a small set of examples, the sender can communicate new symbols using a solve for x pattern and can then build up the small lexicon of symbols needed for a programming language. The recipient would then only need to build an interpreter that processes these instructions and runs them on a simulated computer. This is how many programming languages, such as Java and Python, operate.

This is an area where computer scientists and hobbyists will be able to make significant contributions to the analysis and comprehension effort. While the number of basic instructions needed to build a programming language is small, they can be combined in many ways. The sender might opt for the most minimal language possible, or they might use a larger instruction set to define something that is easier to learn via solve for x examples.

Boolean Arithmetic

Boolean arithmetic is a basic feature in computer programming. These functions can also be implemented in hardware, where they are referred to as *logic gates*. Modern computers consist of billions of these logic gates.

NAND (NOT + AND)

The NAND (NOT AND) operation is a universal logic gate or function, meaning that all other logic functions can be built by chaining NAND operations or gates together. This function accepts two or more inputs and only returns False (0) if all inputs are True (1).

Table 9.1 NAND gate truth table

A	B	Output
0	0	1
0	1	1
1	0	1
1	1	0

Table 9.2 AND gate truth table

A	B	Output
0	0	0
0	1	0
1	0	0
1	1	1

Table 9.3 NOR gate truth table

A	B	Output
0	0	1
0	1	0
1	0	0
1	1	0

Table 9.4 OR gate truth table

A	B	Output
0	0	0
0	1	1
1	0	1
1	1	1

AND

The AND function or gate accepts two or more inputs and returns True (1) only when all of the inputs are True (1).

NOR (NOT OR)

The NOR function or gate, like NAND, is a universal logic function from which other functions can be derived by chaining these together.

OR

The OR function or gate returns True (1) when any or all of its inputs are True (1). It only returns False (0) if all of the inputs are False (0).

NOT

The NOT function or gate returns the inverse of its input, so it returns False (0) when the input is True (1) and vice versa.

Math Operations

A computer is essentially a programmable calculator. Hence, math operations are a fundamental element of computer programs. In fact, you can think of a programming language as a special type of mathematical language that has been extended to include conditional logic and memory, which we will discuss shortly. A typical programming language will define a small set of math operations from which more complex math functions can be built. This will usually include addition, subtraction, multiplication, and division and may include additional functions such as the modulo or remainder. Higher-level math functions such as sine, cosine, and x^y are typically defined in functions composed of lower-level operations (see Modularity and Reuse).

Memory (Variables)

Computer programs often store information in memory so that it may be recalled for future use in other operations and also to share information and coordinate activity with other programs. Low-level programming languages, such as machine language or assembly language, allow programs to directly read and write to RAM (random access memory). This is difficult to follow, so most programming languages provide access to memory via "variables," as shown in the simple example below.

```
# assign the value 3.14159 to a variable named 'pi'
pi = 3.14159
# assign the value 2.5 to a variable named 'radius'
radius = 2.5
# assign the value of pi times radius squared to a
variable 'area'
area = pi * radius * radius
# print the value of the variable named 'area'
print area
```

In the example above, the fragment of code uses three variables and calculates the area of a circle. The variables are used to store the values for pi (a fixed constant), the radius of the circle, and the computed area. The important thing to note here is that the author of the program does not need to worry about where the contents of these variables are stored in memory hardware; this is all handled dynamically by the interpreter that runs the program.

If you are having trouble following this, think of the blackboard in a classroom. A student is tasked with calculating the area of a circle. The value for pi is written down on one panel of the blackboard, while the radius of the circle is written on another panel. The student then calculates the area of the circle by multiplying pi times the square of the radius and writes this value down on a third panel of the blackboard. This is what a computer program does when it reads and writes to memory – the only difference is that it may be manipulating millions of bits of information instead of writing on a chalkboard.

In the case of programs that are designed to run in an interpreted environment (a virtual computer), memory can be addressed in a number of ways, by referencing numeric addresses or named variables that are easier to remember. The address space could be one-dimensional, or it could be an n-dimensional space, to allow for a large and elaborately structured shared memory for programs to use. Programs may also use different mechanisms to access shared memory (accessible to all programs at runtime) and local memory (only accessible to a single program or function within a program). There are many ways this could be done, and no particular architecture is inherently better than others, so it may take time to understand how memory is being used in an alien computing paradigm.

Inter-Program Communication and Input/Output Interfaces

Besides being used to store information for future use, variables and memory can be used to facilitate communication between different programs and also to create input/output interfaces with the external environment.

Inter-Program Communication

Shared memory or variables can be used to facilitate communication between different programs running on the same computer. Let's imagine that we have a few programs running that exchange information with each other. All of the programs can read and write to a shared set of memory locations.

A good analogy is to imagine a classroom with a group of students and a blackboard. Each student is assigned a separate problem to solve. When each student has finished their work, they each write their solution on the blackboard. Some of the problems assigned to students require them to use the solutions from other students' problems as part of their solution. This is similar to how computer programs can use shared memory or variables to coordinate work across multiple programs. There is more to it than that, but this analogy gets the basic idea across.

Now let's consider an example where thousands of programs are running in parallel to run a weather simulation. Each program models a small cube of space in the weather system and communicates with its six adjacent neighbors via shared variables. This allows the output of the simulation in one block of space to be used as an input to the simulations running in neighboring spaces.

When the simulation completes its work for its block of space, it writes the updated temperature, humidity, and air pressure values to variables that the neighboring simulations can read from. Simulations like climate models may consist of millions of relatively simple programs running in parallel, each modeling a small part of a much larger and more complex system.

Serial Versus Parallel Computing

Serial or *sequential computing systems* execute a series of instructions from start to finish. For example, let's imagine that you give a clerical worker a stack of resumes that need to be sorted alphabetically by the applicant's last name.

The worker has a long table to place the sorted resumes on, with A's on one end of the table and Z's on the other. If there is only one worker doing the sorting, this is an example of a serial or sequential task.

This type of task can also be broken down and given to multiple workers, so the same work can be done in a shorter amount of time. Each worker is given a random subset from the stack of resumes and follows the same rules in arranging them on the sorting table. When the last worker is finished sorting his or her stack of resumes, the task is complete. This is an example of *parallel computing*.

Which computing method is best? It depends on the type of problem being solved, and whether it lends itself to being broken down into smaller tasks that can be done in parallel.

Some tasks, such as computing an output based on a small set of inputs, are best done as sequential tasks. Other tasks lend themselves to parallel computing. This is especially true in situations where you have a large number of agents that execute a simple series of steps and share their results with each other. Simulations, such as weather or climate models, work by having many programs running in parallel, each one simulating the behavior of a small part of the larger world being modeled. Artificial neural networks are another example of algorithms that lend themselves to parallelism.

Input/Output Interfaces

Many programs are designed to interact with their users. For this, it is necessary to create a mechanism for the user to enter information into the program and for the program to send or display information to the user. This is known as an *input/output interface.*

One way to do this is to set aside special variables or blocks of memory for this purpose. This is how computer graphics displays work. They allocate a block of memory that contains the current values of pixels to be displayed. To update the display, a program writes to the memory address that corresponds to the pixel to be updated. The video display adapter reads from this same block of memory and repaints the display with the updated information.

Let's consider how this might work in a virtual, or simulated, computer. Any program or function that generates a visual output would always write to special regions in virtual memory. The operator of the virtual computer will have a "God's eye" view of the system and will be able to inspect the flow of information within the system, including what memory locations are in use and what they contain. If we see that certain memory locations are accessed by many programs, that may be a hint that it is being used as an interface.

A video display interface is a good example of an output interface, but what about inputs used to feed into a program? Let's imagine the sender wants to include a Pong-like video game.

This game has three inputs and two outputs:

- Memory location 1 – the position of the paddle on the left side of the screen (input interface: a few bits)
- Memory location 2 – the position of the paddle on the right side of the screen (input interface: a few bits)
- Memory location 3 – start game/reset game switch (input interface: one bit)
- Memory location 4 – video memory for the display (output interface: about a million bits; exactly how many depends on the display resolution)
- Memory location 5 – output audio for the game (output interface: a few thousand bits)

The program that implements the game will read memory locations 1 and 2 for information about the left and right paddle positions and memory location 3 to see if the user wants to restart the game. The program calculates the next position of the ball and then writes to memory location 4 to update the display and writes to memory location 5 to generate audio (e.g., a beep upon paddle-ball collision). Then rinse and repeat ad infinitum.

This pattern can be used to create any number of input and output interfaces for programs running on a virtual computer. These I/O interfaces may be difficult to identify at first, but they will enable programs to interact with the external environment (or a simulated external environment).

Now what's interesting about this is that even though this program runs on an imaginary computer with imaginary memory registers, there is nothing stopping the imaginary computer from passing this information on to a real-world interface so that people can interact with the program.

Comparison Operations and Branching

Computer programs contain *conditional logic*, where they execute different instructions depending on the value of a variable or test condition. The program first compares the value of an item to a test condition (equal to, greater than, less than, etc.) and, based on that comparison, follows a different path of instructions.

A classic example of this pattern is the IF THEN ELSE statement. This takes the form:

```
if a > 0:
            print "a is a positive number"
        else-if a < 0:
              print "a is a negative number"
        else-if a == 0:
              print "a is zero"
        else:
            print "a is not a number"
```

Looping is another form of branching, where a computer will continue executing a block of instructions until a test condition is met. The WHILE loop is an example of this pattern and takes the form:

```
x = 1
        while x <= 100:
              print x
              x = x + 1
```

The sample program above counts from 1 to 100. The basic pattern used in a WHILE loop is to continue doing something while a specific condition is met and then stop when it is no longer met.

Different programming languages implement these patterns in varying ways, but they all have them, as they allow a complex series of instructions to be expressed in a compact form.

Modularity and Reuse

An extensible programming language should also enable its users to build higher-level functions that can be reused and incorporated into other functions. The function to calculate the sine of an angle is a good example, as it can be calculated through a combination of simple arithmetic and conditional branching by using the *Taylor series*, one of the ways to estimate the sine of an angle, as shown below:

$$sine(a) \approx \frac{a^1}{1} - \frac{a^3}{3!} + \frac{a^5}{5!} - \frac{a^7}{7!} \cdots$$

in expanded form

$$sine(a) \approx \frac{a}{1} - \frac{a \times a \times a}{3 \times 2 \times 1} + \frac{a \times a \times a \times a \times a}{5 \times 4 \times 3 \times 2 \times 1} - \frac{a \times a \times a \times a \times a \times a \times a}{7 \times 6 \times 5 \times 4 \times 3 \times 2 \times 1} \cdots$$

This sequence of instructions can be expressed in Python, a widely used programming language, with the code below. Note that we assume the functions power() and factorial() have been defined previously.

```
def sine(a, steps = 9):
        a = a % (2 * 3.14159265359)
        v = 0
        n = 1
        p = 1
        while n < (steps * 2):
                if p == 1:
                        v = v + (power(a, n)/factorial(n))
                        p = -1
                else:
                        v = v - (power(a, n)/factorial(n))
                        p = 1
                n = n + 2
```

```
return v
```

The function above calculates the approximation for the sine of the angle a, where a is given in radians, and steps is the number of steps in the Taylor series. If omitted, it defaults to 9 steps or levels of precision. Once defined, the user can use a compact expression like this one whenever they need to calculate the sine for an angle.

$$y = sine(a)$$

Modularity eliminates the need to repeat the instructions for a function every time it is used. This reduces the size of the programs, which is vital given the constraints in interstellar communication. This also enables users to create higher-level functions using reusable functions as building blocks, all of which reduce to the small lexicon of math and logic operations used to define the foundation of the language. It is also interesting to note how this approach can be used to define more complex math operations and symbols to be associated with them via reusable functions.

Applying These Concepts to an ET Programming Language

It is important to note that the sender could combine these patterns in any number of ways and that the examples provided are in human-readable form to aid readers of this book. An alien author would be very unlikely to have any knowledge of human languages, much less English! While we shouldn't expect anything like BASIC or Python, an ET programming language would probably share similar math and logic functions.

Instead of using English notation to describe the language's basic operations and functions derived from them, it would make sense to use a numeric

Table 9.5 A list of sample functions and their uses

Function number/ ID	English function name	Remarks
2048	Factorial	Computes the factorial of an integer
2049	Sine	Computes the sine of an angle in radians
2050	Cosine	Computes the cosine of an angle in radians
2051	Tangent	Computes the tangent of an angle in radians

address space to refer to them. Each reusable function would be assigned a unique numeric address so it is not confused with other functions. We will be able to build tools that enable us to map numeric function references to human-friendly names. Let's say, for example, the programming language defines functions numbered 2049, 2050, and 2051 to calculate the basic trigonometric functions sine, cosine, and tangent. On first encountering references to these functions, we would not know what they are, but by inspecting the outputs they generate in response to their inputs, or by inspecting their underlying code, we could learn their purpose and build a translation table to map numeric labels to human-language labels. Editing tools could automatically perform these conversions, making it easier for us to read and understand these programs.

Lossless Compression Algorithms

Algorithms will be very useful in applying error correction and compression to the data stream, both to make it more resistant to transmission errors and to maximize the amount of information that can be sent over a transmission channel. Let's take a look at how compression algorithms work.

The basic goal of a compression algorithm is to minimize the amount of repetition in a data set. This increases the amount of useful information that can be sent across a communication channel, and it can be done in a number of ways.

The way these programs typically function is to scan through a block of data to find repeating sequences and then rank the sequences by how often they appear and how long they are. The program then assigns codewords to each sequence of symbols, using the shortest codewords for the sequences that appear most often and longer codewords for the sequences that appear less often.

The program then transforms the original data into a sequence of binary codewords and bundles this with a dictionary of the original sequences and the codewords they were mapped to. A decompression algorithm can then read the compressed file and can transform the codewords back into the original data stream using the dictionary. This operation reproduces the original data stream with no loss of information, hence the term *lossless compression*.

Compressors like this can operate near the theoretical limits of a channel's information carrying capacity, yet do not need to know anything about the content they are compressing and decompressing, since both data sets are just a series of numbers or symbol states.

Lossy Decompression Algorithms

Images, video, and audio content can require large amounts of information to represent, something we will discuss in Chap. 10 Images. Lossless compression can reduce this somewhat, but usually not by very much. A 20–25% reduction is considered to be pretty good. *Lossy compression* works by discarding information that can be omitted without being noticed. For example, a typical landscape scene will often contain a mixture of foreground terrain that has a lot of detail and background sky that has little or no detail. A lossy compression algorithm can encode the image based on how much detail is needed to reproduce each part of the image, which can result in an order of magnitude reduction in the amount of information needed to reproduce the image.

The challenge with these compression algorithms is that their inner workings are difficult to guess. Even if you displayed a compressed data set alongside the image it represents, it would not be clear to the viewer what steps were used to transform the compressed data into an image. If the sender can include the algorithm that is used to do this transformation or decompression, the required steps can be arbitrarily complex and require a lot of computation, but

Fig. 9.1 An uncompressed bitmap image (no lossy compression applied). Size: 3.9 megabytes (31.2 megabits). (Image credit: Brian McConnell, Pacifica CA (2020))

Fig. 9.2 A bitmap image with lossy compression (JPEG, with high compression settings). Size: 34.4 kilobytes (275,200 kilobits). Notice that fine detail is blurred and blocky-looking when zoomed in. (Image credit: Brian McConnell)

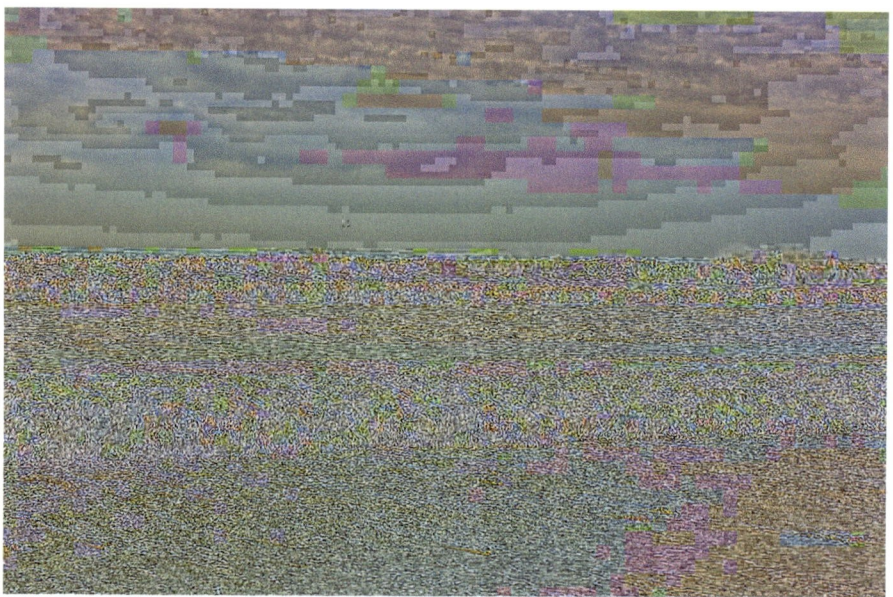

Fig. 9.3 This image displays the difference between the uncompressed and compressed versions of the original image. Notice that information has been deleted from throughout the image, especially in the high detail area in the foreground. The compressed image requires roughly 1/100th the amount of information to represent compared to the uncompressed original, an impressive reduction. (Image credit: Brian McConnell)

that's okay. All the recipient needs to do is to pass the compressed data through this algorithm, much like displaying a JPEG image using a JPEG viewer.

There is no free lunch with lossy compression. The cost of reducing an image or sound file's information footprint is the loss of information, so the reproduced copy is similar but not identical to the original. In most cases, this should be okay, although this should not be applied to scientific data or other collections where the reproduced copy should be identical to the original.

To understand how lossy compression can affect image quality, let's look at an example.

While the uncompressed and compressed images appear similar, on closer inspection it is apparent that some details are blurred or lost in the compressed version of the image. We can make this clearer by displaying an image that represents the difference between the two pictures.

Logic Gate Networks

Another approach toward defining an algorithmic message is to define the computing hardware or logic circuitry on which a program will run and then set the initial states of the circuit. This is similar to the way electrical engineers design and simulate the behavior of complex electronic circuits using programs like MATLAB Algorithmic Communication (2002). If you don't have

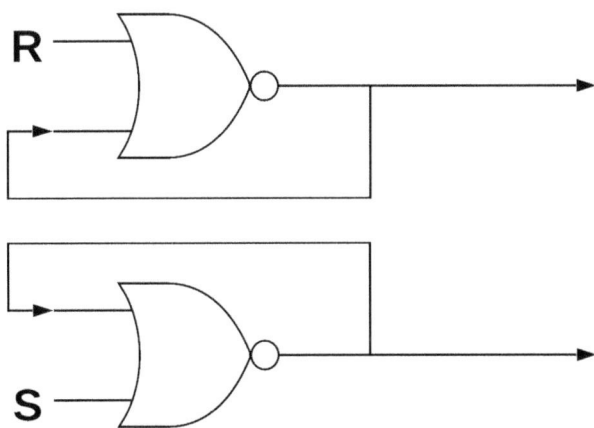

Fig. 9.4 An SR latch, or flip-flop circuit, is created by cross-coupling two NOR gates. The output of the circuit depends not only on the current inputs but also on the current state of the outputs and therefore previous inputs. (Image credit: Brian McConnell)

a background in electronics design and get stuck on this subject, think of this as sending a computer program along with the schematics for the computer it is designed to run on.

This can be done by describing a matrix of universal logic gates such as NAND gates or NOR gates, which can be chained together to create any higher order logic element, which in turn can be chained together to perform higher-level operations. This approach may be appealing if the sender wants to describe a specific type of computing system. The sender would be describing the schematic for a logic circuit such as a central processing unit or memory array.

One advantage of this approach is that it defines the computing hardware needed to implement a given procedure. This is appealing if the sender wants to describe the process in explicit detail or assist the recipient in building hardware that can implement the process very efficiently. One could describe a forward error correction system as a network of logic gates that accept A bits of data as inputs and produce B bits of data as output. This would eliminate any guesswork about how the forward error correction algorithm operates.

The sender could assist the receiver in understanding how to parse logic gate matrices by starting out with simple logic circuits, such as the SR latch depicted in Fig. 8.4, before defining larger and more complex circuits. Both simple and complex circuits can be built up as a set of connections on a virtual circuit board.

Both approaches are valid. Which is best depends on the type of program the sender wishes to describe. If the sender wants to describe a simple game like tic-tac-toe, this can be done with a language that runs on a virtual machine, as the rules of the game are abstract and are not tied to a specific piece of hardware. A complex simulation that is designed to run on many processors in parallel, something like a large neural network, may not lend itself to a procedural programming language and therefore requires its computing substrate to be described in detail.

Another possibility is that logic gate matrices could be generated as output from another program. The sender might want to generate a very large and highly parallel computing system to host something like an artificial neural network. This type of computing system might consist of millions or billions of individually simple computing and memory elements. It would be wasteful to describe this system in long form when a simple program could generate the plan for the computing network with a few nested WHILE loops or their equivalents.

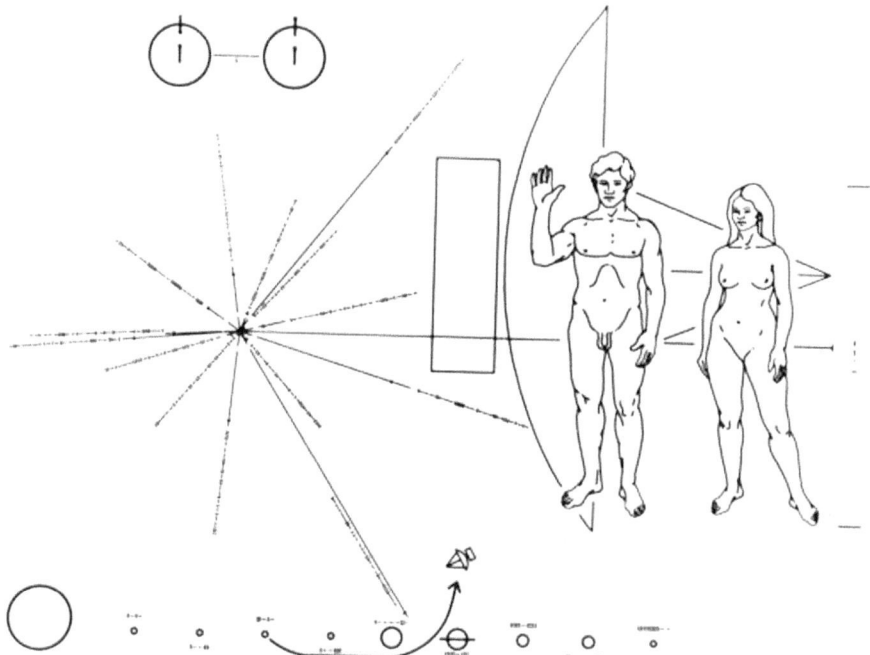

Fig. 9.5 The plaque attached to the Pioneer spacecraft was one of humanity's first attempts to communicate with anyone who might discover a spacecraft in the distant future. It is an example of static information. (Image credit: NASA/Jet Propulsion Laboratory)

The Limits of Computation, AI, and Sentient Messages

Even if the algorithms within an ET message are relatively dumb, they'll still be able to do a lot of interesting things. We know this is possible because of our own experience with computing. Even the most complex programs reduce to a simple set of math and logic operations.

What is perhaps most interesting is that the transmission itself could be a form of intelligence and could even be conscious. This is speculative, but the trends in machine learning point toward some form of artificial intelligence becoming reality. Even if it is confined to relatively narrow domains like image classification, this could have many applications within an interstellar message. An intelligent message would be able to interact with the receiver locally, a nice workaround to the time delays in two-way communication across interstellar distances.

How intelligent could such a system be, given the constraints on information carrying capacity and our own relative inexperience with machine learning?

An interstellar message can be classified in terms of sentience based on a combination of criteria, ranging from a purely static message that does not interact with the recipient to an algorithmic message that is capable of learning from input from the recipient and engaging in two-way communication.

Level 0: Static (Non-interactive) Information

At the low end of the scale, we have purely static information, which we will call Level 0 information. This type of message may contain a variety of media types, such as images, audio, three-dimensional models, and semantic networks, but the message itself is static, just as a book is static, and can only be read by its viewer. This is the type of message most frequently depicted in contact scenarios. This type of message will have a low risk level, except for the possibility of abuse or misinterpretation. This risk should probably not be underestimated considering that human history is replete with bad outcomes from the game of "telephone," especially as it relates to metaphysical or religious topics.

A static message, while it does not interact with the recipient, may contain many different media types. The main limitation of this type of message is that the recipient can only read it and cannot interact with it locally.

Level 1: Intelligently Designed Algorithmic Messages

This type of message may combine static content along with pre-programmed (non-learning) algorithms. This is the type of computer program we are most familiar with, and it is written to fulfill a specific purpose, such as decompressing a file, performing a set of calculations, implementing a game, and so on. This type of program does not learn from its user, and while it may be able to emulate a complex system, it is not sentient in any meaningful sense. Yet even non-sentient programs like this will have useful applications in an interstellar message.

As we discussed earlier, an interstellar communication system will need to be designed so that the receiver can reconstruct information received out of sequence and can also detect and correct transmission gaps or errors without requesting retransmission of the lost data. We showed a simple example of how this could be done by dividing a large block of data into smaller segments and then by retransmitting these periodically. This approach, although straightforward, is also wasteful and reduces the amount of information the link can transmit.

One of the great advantages of incorporating a programming language into the transmission is that the sender can include programs that implement sophisticated forward error correction algorithms that are much more efficient and that can recover lost information in a wide variety of situations. Advanced error correction systems work by appending error correction codes to the information that is to be protected against loss or corruption. In a well-designed system, the sender can also adjust the amount of error correction metadata to compensate for an expected amount of data loss.

As discussed, compression/decompression algorithms (*codecs*) are another example where embedded programs will be useful. Images, audio, and other media types require a lot of information to represent in uncompressed form. Well-designed compression algorithms can reduce the amount of information required to describe an image or audio stream, sometimes by orders of magnitude. While these lossy compression algorithms do not reproduce a perfect replica of the original, they produce a close approximation of the original while using far less data.

These compression algorithms can be quite complex, so it would be difficult for someone to reverse engineer a decompression algorithm just by looking at a compressed data set. That's the neat thing about programs. While they

A Universal Unpack Function

One of the most important programs or functions the sender could include is a utility that reads raw data extracted from a communication channel, unpacks, and, if possible, applies decompression and error correction algorithms to it. It then returns unpacked and error-corrected data as its output. This is similar to the low-level software used on every computer to read data off storage media, apply error correction to it, and present it to the user as an organized collection of data.

A function like this would perform a series of instructions like the following:

- Check to see if the data passed into the function has already been processed and unpacked. If yes, just return the data that was passed in (no further work required).
- If not, is the block of data a valid data element or segment? If not, return a null result and/or an error code.
- If yes, proceed to unpack compressed data using a lossless decompression algorithm, as well as to perform forward error correction to detect and fix corrupted data.

- Was everything unpacked and error-corrected successfully? If yes, return the unpacked data. If no, return a null result or error code.

The benefit of this approach is that the recipient would not need to understand the details of how the function processes data. Once they have figured out the basic instruction set used to run programs, they can rely on this program to faithfully execute whatever sequence of computations and decisions are needed to implement these processing and error correction steps. This would enable the sender to apply sophisticated error correction codes to the transmitted data that would enable the receiver to detect and repair most transmission errors and coverage gaps.

A function like this would also enable the sender to alternate between uncompressed and easily parsed data and optimally encoded data in their transmission and to do so in a way that maximizes the odds that the recipient will be successful in partially comprehending its contents. The sender might send most data segments in plain text on the slow, easily recognized channels that are likely to be noticed first but send most data in a compressed format on the lower power side channels. The recipient would notice that the first-order entropy of the data streams alternates between high and low values, where the low values point to plaintext regions of the transmission.

reduce down to a handful of math and logic symbols that are individually easy to understand, they can be combined to implement arbitrarily complex procedures that can be executed perfectly by a computer.

What risks might programs like this present to the receiver? Any of the programs we discover in an ET message would not be designed to run directly on our computer hardware, but rather would be run on a virtual machine or simulated computer. This will enable us to study and probe the behavior of these programs in a simulated environment that is walled off from other computers and networks.

Level 2: Machine Learning/Specialist AIs

We are already seeing a proliferation of narrow or specialist AI systems that are being used successfully for tasks like speech recognition, machine vision, and language translation. These systems are not computer programs in the sense that they execute sequential instructions that were written by a programmer. Instead, they are modeled after animal nervous systems and learn by observing large sets of examples. Some of these are supervised systems, where the examples are accompanied by the equivalent of cue cards that tell the program what is in the data set it is looking at. The program is left to its own devices to learn how to build these associations. Unsupervised systems learn to recognize patterns without any external intervention.

This class of AI, while not sentient in the sense of being conscious, could be embedded in an interstellar message to answer queries from its receiver. Going in the other direction, an image classification AI could ingest a large number of images or other media samples from its receiver and build its own representation of the information we provide to it. Any civilization we are likely to come in contact with will probably have greater experience with computing and communication, so narrow AI seems like a low bar for them to cross and something we should anticipate.

Level 3: General AI (Sentient Messages)

We may also find that narrow AIs are networked to create a higher-level, general AI. In this design pattern, many specialist AIs are interconnected, each performing a specialist task, such as recognizing edges in an image, while others recognize color or sound patterns.

General artificial intelligence is thought by some people to be an emergent property of such a networked system of specialist AIs. AI pioneer Marvin Minsky describes this pattern in his book *Society of Mind* Minsky marvin (1985, 1986). This idea is also rooted in anatomy, as it is known that certain parts of the brain are focused on different types of function or information, such as the Wernicke's and Broca's areas of the brain, which are involved in the recognition and production of speech

We have yet to invent a true general-purpose artificial intelligence, a system that can match or exceed the flexibility and repertoire of human thought. We only have about 60 years of experience with modern computing systems and have made substantial progress toward intelligent systems in that time. But we don't yet know if the leap from narrow/specialist AIs like image classifiers to general intelligence is a small step a few years in the future or a giant leap that may remain out of our reach for decades or longer.

While we can't predict the applications or behavior of such entities, we can define the capabilities required for an algorithmic system to be considered a general AI, if such a thing is possible.

These will include:

- **A capacity for both supervised and unsupervised learning** – when it observes data sets provided by the sender, it will be able to draw inferences from a limited set of inputs.

- **Theory of mind** – this is a hallmark of intelligence as we define it and refers to the ability to put oneself in another's position, to see or imagine a situation from their perspective.
- **Capacity for nested abstraction** – this refers to the ability to build up a language of symbols that in turn are derived from other symbols. This ability to chain together increasingly abstract symbols allows one to work up from simple literal symbols to symbols that have no direct physical representation but are linked only to abstract ideas.
- **Autonomy and goal-seeking behavior** – the ability to set goals and devise an agenda toward reaching them.

A general AI might have a pretty straightforward intent – to induce its recipients to relay it to additional recipients at new star systems as a way of propagating itself or experiencing new destinations. In this scenario, the AI could be regarded as a form of artificial life that lives within the computing substrates its recipients build for it as it explores the universe.

This possibility is perhaps the most interesting because we only invented computers very recently and are still at an early stage of understanding what the limits of computation are. A technological civilization that is more experienced than us will have explored those limits, and if general AI is possible to build, the message we encounter may itself be a form of sentience.

Finding and Analyzing Algorithmic Systems

Who will be able to participate in the hunt for algorithms in an ET transmission? This is a task that could, and should, involve thousands of people around the world. Anyone with a talent for programming and a hunch to test will be able to contribute to this effort. It is also important to note that many of the important contributions to computer science came from individual contributors or very small teams of people. This is very much unlike Big Science, such as large-scale telescopes like LIGO that require thousands of people to operate them and understand their output.

Many of the most widely used programming languages today were invented by individuals or people who were operating outside of corporate research departments or academia at the time of their inventions. The Python programming language is an excellent example. It was invented in the 1980s by a Dutch programmer, Guido van Rossum, and has since grown to become one of the most widely used programming languages. Others, such as Java and C, were developed by individuals or small teams of people working at corporate research departments.

Cosmic OS

In 2003, artificial intelligence researcher Paul Fitzpatrick, then at the MIT CSAIL lab, developed a complete, worked example of what an algorithmic message from another civilization might look like. The project, called Cosmic OS, builds up a programming language using sets of *solve for x* examples, similar to the processes described in this chapter. His initial goal was to develop a programming language that could run a multi-user role-playing game, as this type of game can be used to explore strategies and game theory.

The project was inspired by Lincos, a language for interstellar discourse that was invented by the Dutch mathematician Hans Freudenthal (1960).The language, rooted in math and logic, built a foundation on these concepts to develop a language that could be used to discuss math and logic in simulated dialogues.

Cosmic OS takes a similar approach by enabling the user to build up a lexicon of logic and programming concepts. From there, it defines a programming language that in turn can simulate scenarios. Readers can try their hands with the project at cosmicos.github.io.

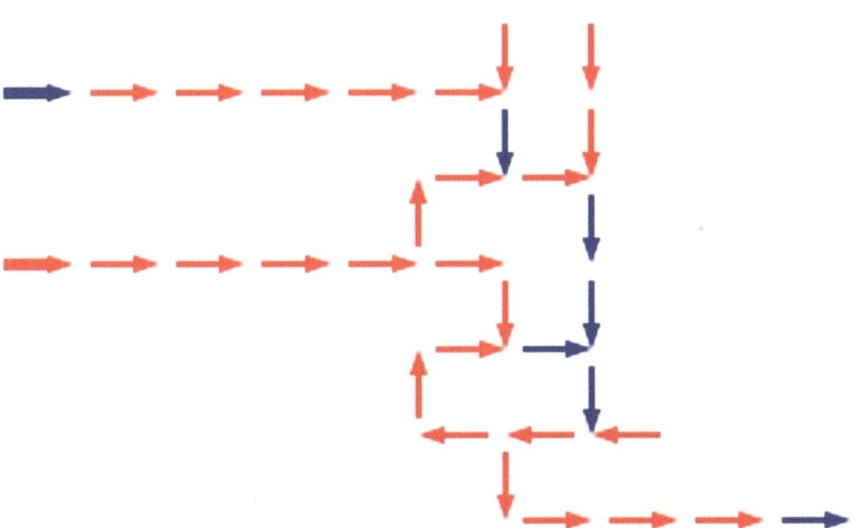

Fig. 9.6 A diagram of a logic circuit defined using T gates, which like NAND gates are universal logic gates that can be combined to create any other logic circuit (AND, OR, etc.). (Image credit: Cosmic OS Github site)

If and when we encounter an ET transmission, we won't have any idea at first what it contains or represents. As previously discussed, there are compelling reasons why such a transmission will include algorithms, yet we won't know how they might be organized and whether they rely on a verbose instruction set or something minimalist, like Paul Fitzpatrick's Cosmic OS[1].

This part of the message analysis and comprehension effort won't require elaborate or expensive infrastructure. Anyone with a computer and Internet access will be able to obtain the data being collected from the transmission and will be able to test their own hypotheses about what it represents. It's possible that the most important insights will come from unexpected people who approach the problem in novel ways.

Checklist of Patterns to Watch for

As part of inspecting the data stream for algorithmic components, we'll want to ask the following questions.

What is the basic unit of information? Is it a raw bitstream? Does it use fixed-length words? Does it use variable-length words, such as we see in the UTF-8 encoding? (Note: the term "word" here refers to organizing a series of digits into larger groupings.)

How is data structured? For example, is there something akin to open and close parentheses that can be used to group a series of numbers into arbitrarily complex n-dimensional data structures?

Is there a mechanism for differentiating between data, operands, and memory addresses?

Is there a mechanism for defining basic data types such as integers, floating point numbers, imaginary numbers, etc.?

Is there an address space or naming convention for reusable functions?

How are variables or memory referenced? Is a one-dimensional or n-dimensional address space used? Is there a notion of global and local scope that governs what memory addresses are accessible and to which programs or functions?

Does the system describe a specific computing substrate via logic gate matrices, or does it describe an interpreted programming language that runs on a virtual computer?

Does the system rely on a minimal instruction set as its foundation or something more verbose? Are the basic operands also mapped into a larger

---

[1] Cosmic OS project website. https://cosmicos.github.io/.

address space (such as a general-purpose semantic network or something like the Unicode address space)?

How do programs read and write from variables/memory at runtime? Are there recognizable patterns (e.g., do most programs appear to write output to a specific address range, which might hint at input/output interfaces)?

Can reserved commands and reusable functions be mapped, at least partially, into human-friendly forms? For example, function 2049() → sine().

Do programs or functions appear to be designed to process data from the raw data stream, for example, to apply forward error correction algorithms to the data stream?

Do programs use shared variables to communicate with each other or trigger actions in each other? If so, how extensive is this communication, and is there a pattern to how information is shared?

Are there programs that appear to be designed to interact with users or the external environment in some way? How sophisticated are these interactions?

Is there a pattern of prolific communication between many programs? This could hint at a large-scale simulation or artificial intelligence.

Is it possible that obviously bad design patterns persisted in spite of their badness – e.g., 500 million years ago an alien Bill Gates forced a bad design decision in "ET-DOS" memory management that persists to this day?

In asking these and other questions, it is important not to fall into the trap of assuming that the paradigms we use in computing are in any way universal. Most are arbitrary and are byproducts of which systems won out in the commercial marketplace as the computing industry developed. This is another reason why it will be good to have a large number of people from diverse backgrounds trying out lots of different approaches to this puzzle and sharing their successes and open source software with the community as they work through this process.

References

Dean Alex (2013), Javascript turbulence simulator, http://neuroid.co.uk/lab/fluid/.

Algorithmic Communication with Extraterrestrial Intelligence, McConnell, B. S., Bioastronomy 2002: Life Among the Stars, Proceedings of IAU Symposium #213. Edited by R. Norris, and F. Stootman. San Francisco: Astronomical Society of the Pacific, 2003., p.445.

Minsky, Marvin. Society of Mind. Touchstone/Simon & Schuster. 1985, 1986.

Freudenthal, Hans (1960). *Lincos, Design of a Language for Cosmic Intercourse.* Amsterdam: North-Holland.

10

Images

Photographs could turn out to be an important and perhaps dominant media type in an ET message. The simplest reason for this, as we have mentioned previously, is that in order for the sender or receiver to have any chance of succeeding at interstellar communication, they will need to understand astronomy, which is largely based on photography.

It is reasonable to assume that both parties will understand photography at least on a technical level and that this will form the basis for a common foundation for communication. It is a particularly important medium because photographs can represent objects and scenes ranging from microscopic to cosmological scales, something no other media type can do. Perhaps more importantly, photographs are a way of enabling someone to remotely view a scene. In the context of interstellar communication, it is almost like teleportation. Imagine receiving panoramic photographs of an alien world, its environs and structures. Even if you had no idea why someone chose to share a particular scene, it would certainly be interesting to see.

Some will argue that this assumption is anthropocentric, and in a sense it is. But note that it is separate from the discussion about whether vision is necessarily the primary sense of another species. It is possible to imagine a species whose primary sense is auditory but has worked out how to map electromagnetic radiation into a form they can understand. This is similar to how we developed ultrasound technology that enables us to transform ultrasonic echoes into images we can understand.

Photographs are also an interesting medium for digital communication because the process used to represent them is quite simple. A digital photograph typically represents a scene as a two-dimensional array of numbers that

B. S. McConnell, *The Alien Communication Handbook*, Astronomers' Universe,
https://doi.org/10.1007/978-3-030-74845-6_10

Fig. 10.1 Part of a panoramic photo of Mars' Gale Crater, taken by the Mars Curiosity rover on Oct 25, 2017. (Source image credit: NASA/Jet Propulsion Laboratory/Mars Exploration Program, mars.nasa.gov (NASA/Jet Propulsion Laboratory. **Wide-Angle Panorama from Ridge In Mars' Gale Crater**. https://www.jpl.nasa.gov/spaceimages/details.php?id=PIA22210))

represent the relative brightness level of each picture element or pixel. The amount of information needed to represent an image is easy to estimate as the number of pixels times the number of bits used for each pixel. A 1000 pixel by 1000 pixel grayscale image with 1024 possible brightness levels, or 10 bits per pixel, will require 10,000,000 bits. Even a slow communication link will be able to transmit many such images per day.

Bitmaps

Bitmap images are typically encoded as two-dimensional arrays of numbers. This format is easy to decode because the viewer only needs to guess a few pieces of information to correctly extract an encoded image. Random noise or incorrectly decoded image data will look like static when rendered as an image. The sender can make the decoding process easier for the receiver by repeating reference images that contain easily recognized features such as those found in planetary images. By guessing just a few parameters, such as the number of bits used to represent the brightness level for a pixel, one can quickly discern bitmap images from other types of data.

The first place we will want to start is by testing different assumptions about the number of bits used to represent each pixel's brightness level. It's good to start here because the number of possibilities is small. Unless the goal is to transmit images with extremely high dynamic range, not that many bits per pixel are needed to yield good results. Generally, 6–16 bits per pixel per color channel is sufficient, so with a few guesses this parameter can be nailed

Fig. 10.2 The "Blue Marble" photograph of the Earth taken by the Apollo 17 astronauts. Planetary images like this have a universal pattern, as they consist of a round object surrounded by a featureless black background. The shading or drop-off in light levels near the edges also follows predictable patterns. These will be excellent reference and calibration images. (Image credit: NASA)

down. With the correct value in place for that, the rest of the image becomes easily visible, even if the other parameters are not quite right. Guess wrong, and you'll see static.

The next step is to test different assumptions about the dimensions of the array (number of horizontal and vertical positions and scan direction). Figuring out the X by Y dimensions of a two-dimensional image will be straightforward, especially if the pictures contain recognizable scenes, such as terrain, planetary photos, or manufactured structures. The reason for this is simply that if either of these numbers is off, the resulting image will be slanted in one direction or the other. A sender who is attempting to initiate communication will be wise to include a frequently repeated reference photo, such as space photo similar to the photos taken by the Apollo 17 crew. The image of a planet or moon against empty space will be recognizable to any astronomer and can also be used to include cues for fine calibration.

Figure 10.3 shows a bitmap image of a planet embedded in a larger data set that appears to be random. This is a bitstream visualization where each bit in the data stream is either a 1 (white) or 0 (black). Notice that the elliptical shape of the object is immediately obvious.

This also provides a clue about how many bits are used to represent each pixel, as the round object is stretched 8:1 horizontally, a hint that each pixel is represented by 8 bits. The vertical lines in the image region also hint at this. The receiver then only needs to guess whether the bits are sent in most to least significant or least to most significant order and then will be able to extract the original image.

The displayed light levels will also need to be calibrated. An 8-bit pixel can have a value ranging from 0 (darkest) to 255 (brightest), which means nothing in terms of physical units. This has to be translated into an actual display brightness, and the brightness curve may be nonlinear to allow for greater

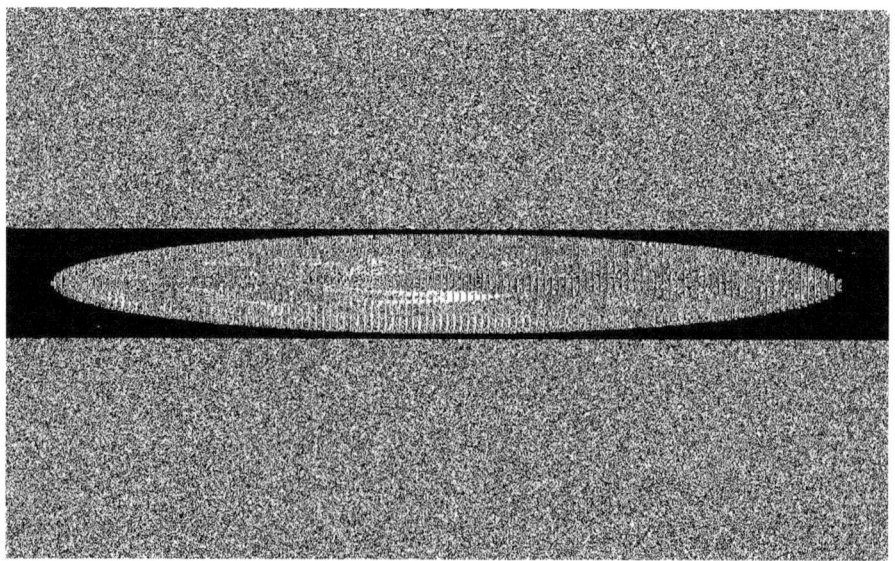

Fig. 10.3 An example of what a planetary image would look like within a bitstream surrounded by a larger set of data that appears to be random. Notice how easy this pattern is to see. (Image credit: Brian McConnell)

dynamic range. This is known as *gamma correction*, where a linear scale is translated into a nonlinear scale as follows:

$$V_{out} = V^{\gamma}_{in}$$

In most display systems designed for humans, the gamma factor is typically set at 0.45 for the encoding stage and 2.2 for the decoding stage. This takes advantage of the fact that human vision perceives brightness on a logarithmic scale. This factor is likely to vary by species, so we should not assume that this type of correction factor is universal. This is another example of where reference images such as planetary images will be useful, because the receiver can apply and adjust correction factors like this until they find the best fit with comparable reference images.

So far we have assumed that images are planar and two-dimensional. This arrangement requires the least amount of information to represent, but we should be on the lookout for other formats, such as cylindrical (panoramic) or spherical projections. These images should also be easily recognizable, but will appear to be distorted when displayed in a planar projection.

Encoding Video on the Voyager Golden Records

The Golden Records that flew with the Voyage space probes were designed to include both audio and still pictures. The recording medium was analog and used a stylus to play back the analog signal etched into the record's grooves. The record was made of metal, not vinyl, but otherwise worked much like the LPs of the day.

The records recorded pictures as sets of 512 scan lines, each of which drew out 1 line of the image, similar to how analog television broadcasts worked. This signal, like the audio content, was analog, so the brightness values varied continuously from dark to light. An explanation of how to decode the content of the record was provided on the cover that enclosed the record (an annotated version of the cover is shown below).

The audio from the record was available and could be used to reproduce the images encoded on the records, which Ron Barry did for the 40th anniversary of the launch of the Voyager spacecraft.

The Golden Record was an impressive accomplishment and was ahead of its time in many ways. The probe was launched in 1977, at a time when digital computing was in its infancy. The Voyager spacecraft itself had less than

EXPLANATION OF RECORDING COVER DIAGRAM

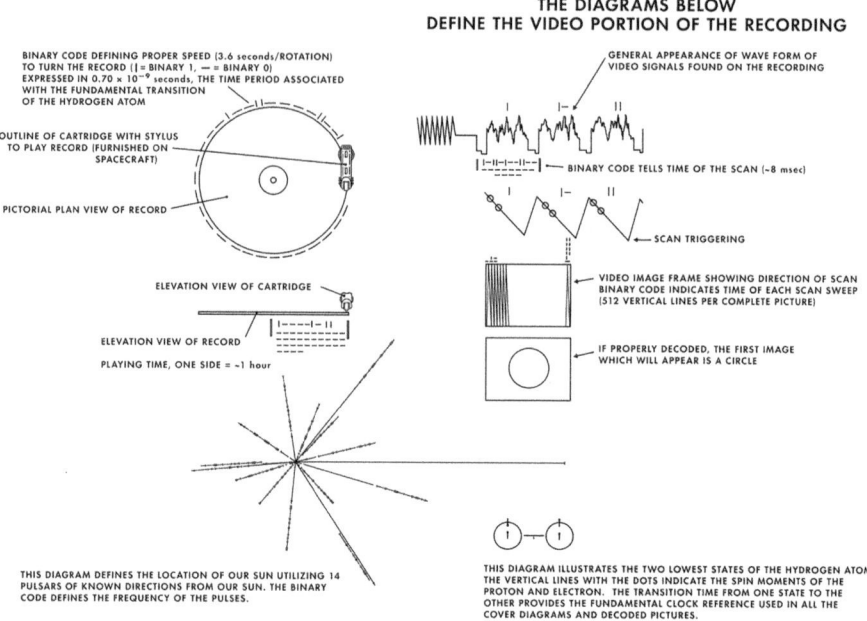

Fig. 10.4 An annotated version of the Golden Record cover. (Image credit: NASA Jet Propulsion Library (NASA/Jet Propulsion Library. **Explanation Of Recording Cover Diagram**. https://voyager.jpl.nasa.gov/golden-record/golden-record-cover/))

100 kilobytes of memory, and digital audio/video recording technology was still years in the future. The record made the best use of the analog technology available at the time and was designed so that a naive receiver would be able to make sense of the information encoded on the record.

Assisting the Receiver in the Decoding Process

The examples discussed so far draw from conventions used in our computing systems. Our notation is not necessarily the only way to encode images. Our conventions are artifacts from which computing systems won in the commercial marketplace years ago, as well as the perceptual capabilities and preferences of human users. There are a number of things the sender can do to help a naive receiver recognize images for what they are. One way to do this is to send several different encodings of a set of images.

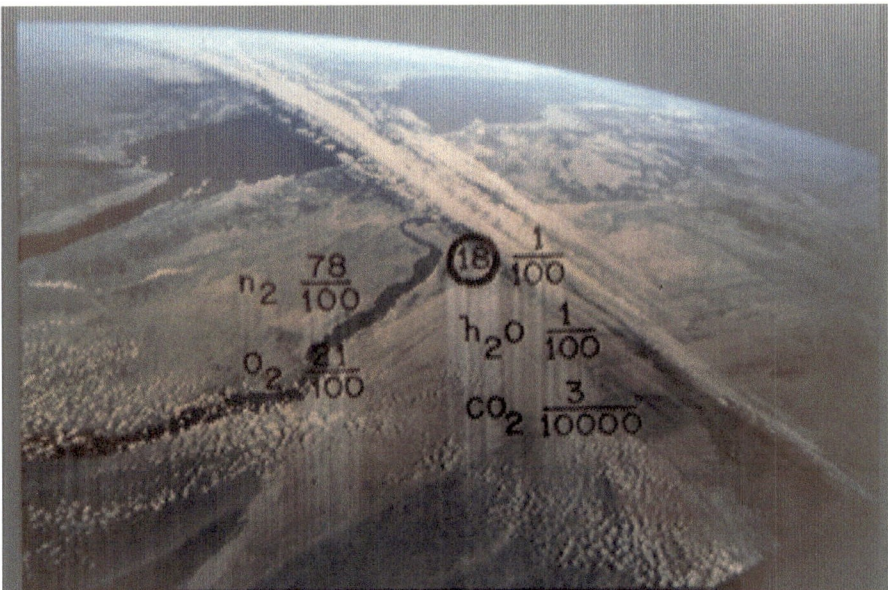

Fig. 10.5 A reconstructed color image, decoded from audio data from the Golden Record. (Image credit: Ron Barry (How To Decode Images On The Voyager Golden Record", Ron Barry, Boing Boing, Sept fifth 2017, https://boingboing.net/2017/09/05/how-to-decode-the-images-on-th.html)

Let's say that the sender wants to teach the receiver to recognize several different image encodings, ranging from an extremely easy but inefficient encoding to an efficient and highly compressed encoding. They could do this by sending several different representations of the same data collection. Each example might have the same collection number, similar to the filename on a computer, while another identifier is varied with each representation. This is one way of hinting that each is a different representation of the same thing.

Using the Blue Marble photo from Apollo 17 as an example, we can anticipate what this sort of pattern would look like. One encoding represents each pixel in the image as a 16 x 16 grid of black or white dots that, added up, translate to 256 different light levels. This is a very inefficient encoding, as it requires 256 bits for each pixel. While it is inefficient, it is trivially easy to figure out the encoding, as the recipient simply needs to guess the number of pixels per row. This process, called *dithering*, is how most newspaper printing processes render photographs.

A bitmap can also be built up from a stack of bitplanes, where each plane is a two-dimensional array of binary digits that can be either a 1 or a 0. Let's say that you want to encode an image that has 256 possible light levels

Fig. 10.6 The Blue Marble photo, taken by the Apollo 17 astronauts; each pixel is represented by a 16 x 16 grid of black or white (1 bit) pixels. This raw bit field allows for 256 discrete luminance values. It is a very inefficient encoding at 256 bits per pixel and requires 32 times as much data compared to a conventional bitmap (8 bits per pixel), but the bit field is easy to decode. (Source image credit: NASA)

ranging from 0 for black to 255 for white. This works out to 2^8 possible values and can be represented with 8 bitplanes.

The first image in the sequence is for the most significant digit, where a 1 maps to 128 and a 0 maps to 0. The second image in the sequence is for the next most significant digit, where a 1 maps to 64 and a 0 maps to 0 and so on. Notice that the major features of the image are evident in the most significant bitplanes, while fine shading detail is encoded in the less significant bitplanes, which appear to be full of noise.

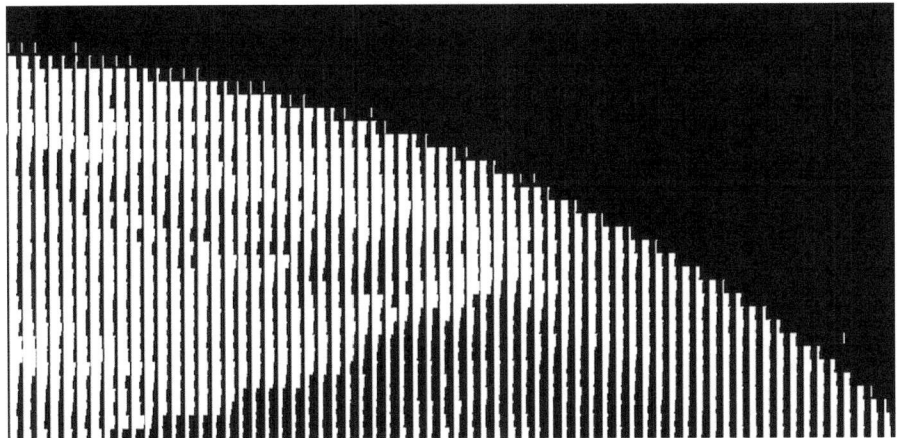

Fig. 10.7 A zoomed in view of the bit field. Notice that the 16 x 16 grid pattern becomes evident. (Image credit: Brian McConnell)

Fig. 10.8 The Blue Marble photo taken by the Apollo 17 astronauts, displayed in a bitplane encoding. Each image in the sequence is a different slice through the bitplane, going from the most to least significant digit from left to right. The bottom row of images contains the sum of the bitplanes. Notice how fine shading detail builds up from left to right. (Source image credit: NASA)

The process for decoding images from a bitplane encoding is to calculate the brightness of each pixel in the decoded image by summing the pixels from each bitplane as illustrated in the image sequence below, where the leftmost image only has two brightness values (0 or 128), but as we sum in information from additional bitplanes, finer brightness detail is revealed. In the rightmost decoded image, there are 256 possible brightness levels.

This encoding has a number of advantages, one of which is that an image can be described using whatever number of bitplanes are required to achieve the desired amount of brightness detail in the decoded image. The basic features of an image are usually pretty easy to recognize in the first and second bitplanes, especially for planetary images and space photography. The aspect ratio of the original image is also preserved in each bitplane, which may also assist a receiver in understanding the image encoding.

The next image in the sequence is also efficiently encoded at 9 bits per pixel but still results in a mostly undistorted image when viewed as an unencoded bit field. In this image each pixel is represented by a 3x3 bit array (9 bits). This format preserves the original aspect ratio of the image, even when displayed as a raw bit field. The brightness levels appear distorted, but the contours of the original image are easily recognizable, especially for calibration images such as planetary images.

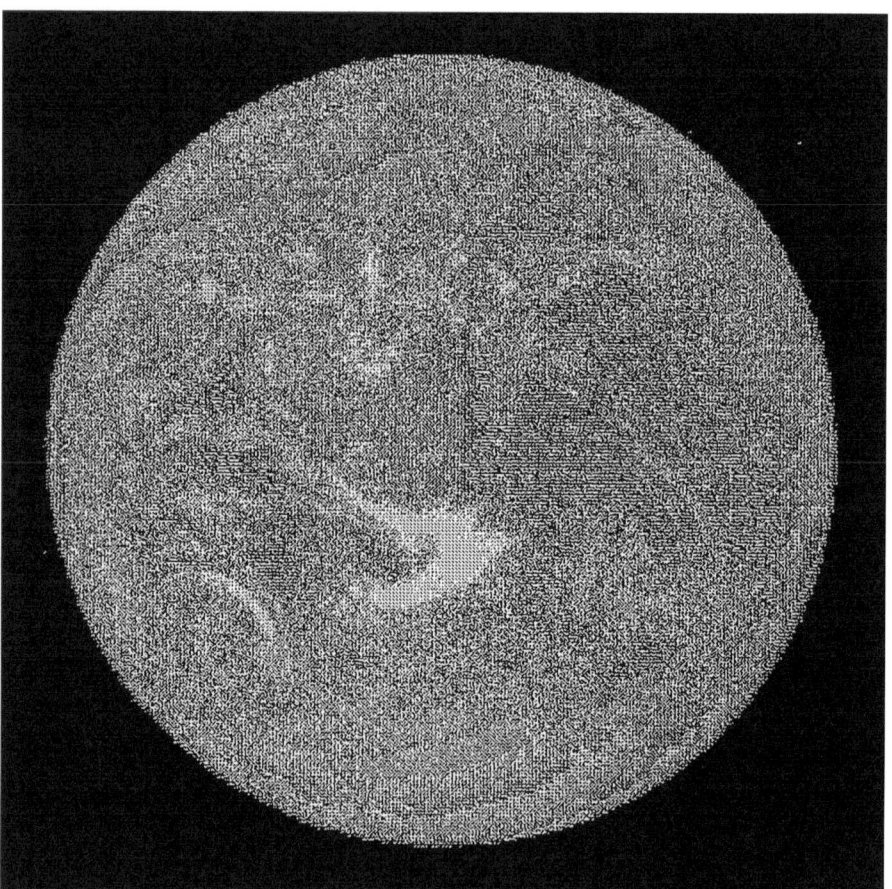

Fig. 10.9 In this bit field, each pixel is represented by a 3x3 array of bits (black = 0, white = 1). In this encoding, the bits are written from left to right, and top to down in each cell, in least to most significant bit order. The aspect ratio of the original image is preserved, but luminance information appears distorted. (Image credit: Brian McConnell)

This encoding is still easy to view in raw form. Once the receiver guesses the row length (256 pixels in this example), the basic contours and prominent features of the image are easily visible. The receiver can validate their guesses at how to decode this image by comparing the brightness value of each decoded pixel against its counterpart from the previous image encoding.

The next image encoding in the sequence is a linear encoding, where the bit values for each pixel are read out sequentially (8 bits per pixel in this example). This results in a more distorted image when it is viewed as a raw bit field. The aspect ratio is stretched horizontally by a factor of 8:1, which also hints at the number of bits used to encode each pixel. Planetary images are still recognizable, but more complex scenes will be difficult to recognize.

The sender might also send one or more compressed versions of the same image and assign different data types to those encodings of the reference image. When the compressed images are viewed as bit fields, they would not be recognizable as images at all, but rather would look like random noise. The receiver would still be able to infer that these encodings produce the same image as the other encodings and would be able to use them to validate their attempts at decoding compressed images. The decompression algorithms could also be embedded elsewhere in the data stream (see Algorithmic Communication Systems), in which case the sender could send most images in a compressed encoding to maximize the number of images that can be sent in a given amount of time.

Notice that what we did here was to send several different representations, or encodings, of the same image. Each encoding is associated with a different data type ID, while the collection ID (akin to a filename) stays the same. The uncompressed image encodings can be inspected visually and should be straightforward to decode. The compressed image encodings have little or no obvious structure when inspected visually, so the receiver will need to figure out how the compression process works, but at least they will be able to infer that these are possibly a type of image encoding.

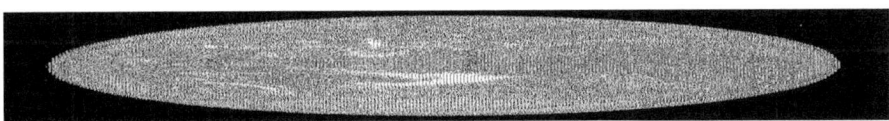

Fig. 10.10   An 8-bit per pixel linear encoding. When viewed as an image, this bit field's aspect ratio is distorted, which will make recognizing some images difficult. Planetary images are still easy to recognize because of the large field of zeros (dark pixels) surrounding an elliptical shape. (Image credit: Brian McConnell)

Fig. 10.11 A bit field display of lossless compression encoding of the Blue Marble image (256 x 256 pixels, 8 bits per pixel, PNG format). Notice that this appears to be nearly random with no visible structure or features, except the topmost rows of the image file (this is where structured header information is found). (Image credit: Brian McConnell)

Color Images

Color conveys additional information that is lost in grayscale images. It is especially important in scientific imaging because color conveys information about the chemical composition of objects. The sender will not know anything about the specifics of human vision, nor can we assume we know about theirs. That said, an astronomically literate civilization will understand the concept of color and its scientific importance.

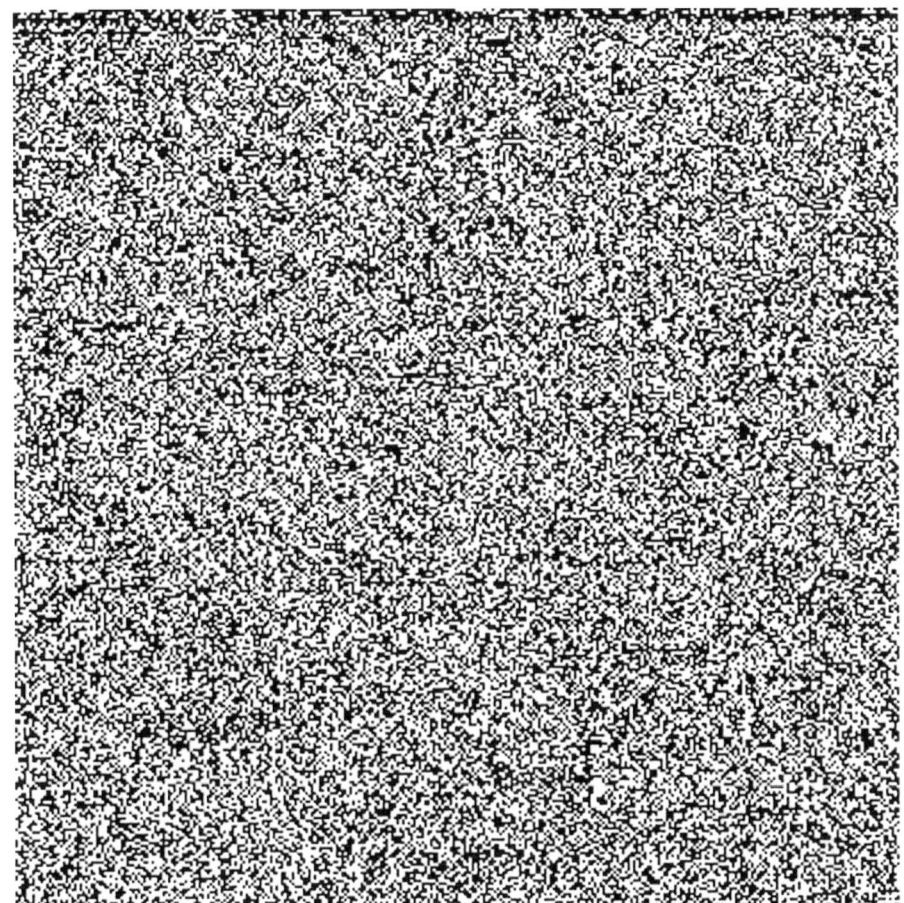

Fig. 10.12 A bit field display of lossy compression encoding of the Blue Marble image (256 x 256 pixels, grayscale, JPEG format). Notice that this appears nearly random. The bit field is also smaller because the lossy compression encoding (a concept that will be discussed shortly) further reduces the number of bits required to represent the image. (Image credit: Brian McConnell)

A good analogy to use is our understanding of infrared and ultraviolet light. Humans can't directly see these colors of light, but we understand what they are, and we know how to build cameras to take pictures in these ranges of color even though we can't see them directly.

So how can one build a system for representing color into bitmaps while retaining the simplicity of monochromatic bitmaps? One way to do this is to simply transmit one image for each color channel, or range of colors, used in an image. Humans are sensitive to light in three main color bands – red, green, and blue – so if we were designing the system around human viewers, we could simply repeat the image once for each of these color bands.

The viewer would see three separate images that appear similar, but not quite identical. A numerical analysis of the three images would reveal the difference between each picture and the others, which would give clues that this is a multicolor channel image. The question then is: how can the sender indicate which color channel is which without having a shared vocabulary for color?

One way the sender can enable the receiver to correctly calibrate color images is to send images of mutually observable astronomical objects. The Cat's Eye Nebula, shown below, is 3300 light-years away, so it would appear virtually identical to a nearby observer. The process of calibrating color reproduction would simply be to compare the images sent in the transmission with our own observations in various color bands to find the closest match.

Those are just a few examples of how this could be done, but it illustrates how the sender can embed information about color in an image by referencing shared examples. Despite the simplicity of the approach, it will enable the sender to include images with whatever level of color detail is desired.

How many color channels are enough? If the information carrying capacity of the communication channel is limited, the sender will be forced to make tradeoffs about which images to send, what resolution to send them at, and also how much color information to provide. Basically they will have to ask, what are the most important features of a specific image? Is it more important to show features in sharp detail, such as the ridges or vegetation in a mountain range? Or is it more important to show color information in greater detail?

A broader question is how much color information will be needed so that receivers with different visual capabilities can transform images into a color system that is compatible with theirs. Humans are sensitive to red, green, and

Fig. 10.13 The Earth, as photographed by the Apollo 17 astronauts in three color bands (red, green, and blue respectively). (Image courtesy of NASA)

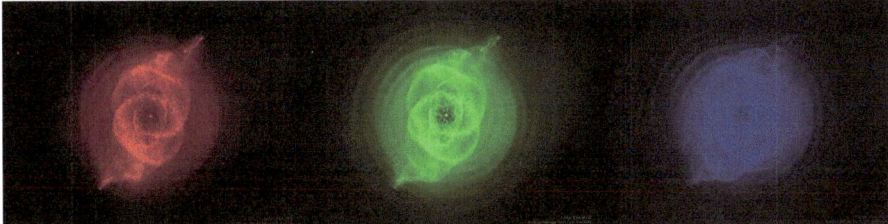

Fig. 10.14 The Cat's Eye Nebula, shown in red, green, and blue color bands respectively. By using mutually observable reference objects, the sender can provide the receiver with another way to calibrate multi-channel color images. (Image credit: NASA/ESA/HEIC/Hubble Heritage Team)

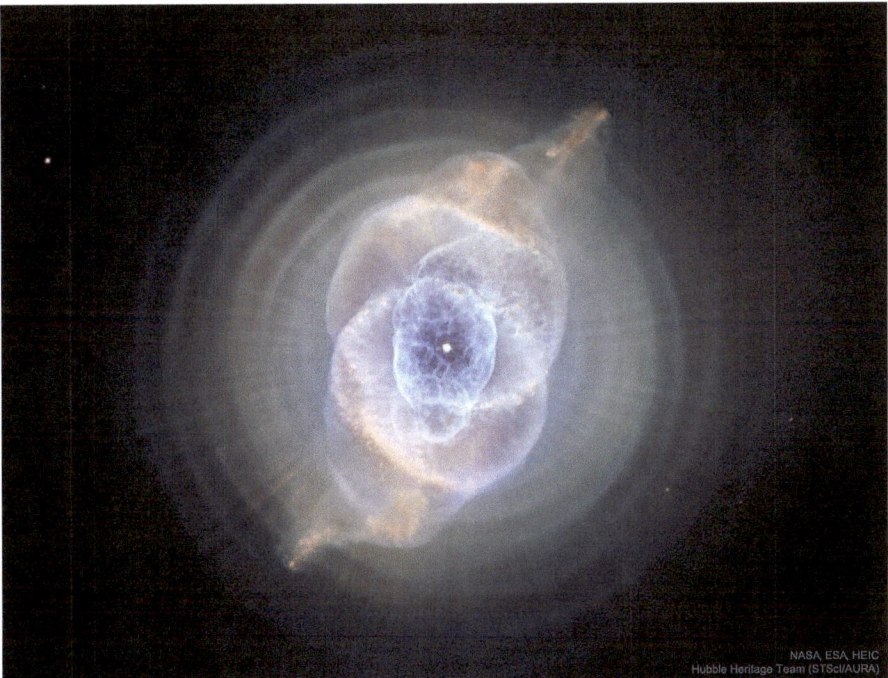

Fig. 10.15 The three images when combined into red, green, and blue color channels yield this full-color image of the Cat's Eye Nebula. Scientists routinely combine color-filtered images from space probes (often different color bands than found in human vision) and transform these into "false color" images that match what the human eye would see. We could use the same technique with multicolor band ET images. (Image credit: NASA/ESA/HEIC/Hubble Heritage Team)

blue, while dogs see primarily in yellow and blue color bands. Many insects can see ultraviolet light, which we cannot, but are blind in longer wavelengths such as reds. So how do you send color images that can accommodate viewers whose sensitivity color may differ so widely? A good approach will be to send images with enough color information that they can be down-converted into reduced color spaces.

Scientific Notation About Color

So far, we have shown how to hint at color by using images of mutually observable objects, such as distant nebulae. This approach should work provided that the receiver is able to take images of the same objects and compare those to the images sent in the transmission. Another pathway to comprehending color information is to use direct scientific references to color information.

If the sender has defined precise units of time and distance elsewhere in the transmission (see the chapter Communicating Fundamental Units and Scientific Information), they can refer to the color using wavelength, which is measured in distance units. Red light has a wavelength of about 680 billionths of a meter (680 nm). This wavelength can be expressed as a multiple of the basic distance unit described in a scientific primer.

Another way a sender can do this is by including solar spectra in calibration images. The light emitted by stars, when split out into a spectrum, includes dark lines where light at specific wavelengths is absorbed by the star's atmosphere. These are also known as Fraunhofer lines. By referencing these, the sender can provide another pathway to understanding color information in images.

We can see that there are at least three different ways to communicate information about color and images, including comparative and explicit methods, to provide the recipient with multiple paths to comprehension.

Fig. 10.16 A representation of the sun's light spectrum in visible colors. This was used as a calibration image on the Golden Records attached to the Voyager space probes. Fraunhofer lines are clearly visible and could be used to work out how to reproduce the color images included in the collection. (Image credit: National Astronomy and Ionosphere Center)

Image Compression

One problem is that high-resolution images require a lot of information to represent. This has long been a problem on terrestrial communication networks, where the solution has been to employ compression algorithms to compact the image files, as we discussed in Chap. 9 on Algorithmic Communication Systems. These algorithms fall into two general categories: lossless and lossy compression. *Lossless compression* allows the original data to be recovered without losing any information, meaning the uncompressed data is identical to the original before compression. *Lossy compression* algorithms throw away unimportant details so the compressed image is similar to but not identical to the original.

The problem with using compression algorithms in an ET message, at least in the parts of the message that are likely to be among the first encountered, is that the receiver needs to know the details of the algorithm used to compress the image. Depending on the sophistication of the process used, this can make the data look like random noise and thus presents an obstacle to intelligibility. That said, there are some approaches that can be used to reduce the amount of information needed and that do not require a sophisticated encoding process.

The first is by curating images wisely, so that important images are sent most frequently or in the highest resolution. The sender controls how many bits can be delivered on any given communication link and so can decide what part of the data budget is allocated to different collections. This requires no additional effort on the part of the receiver, who might also notice that particular images are resent at regular intervals or are sent in higher resolution, so maybe they are especially important. The information footprint of an image varies with image area and the number of color channels used. Reducing the pixel count by 50% in each dimension reduces the information footprint by a factor of 4, so the sender can optimize image size based on what resolution or number of color bands is really needed.

Lossy compression schemes, such as the JPEG format, work by deleting unimportant information from an image. A typical example would be an outdoor photo taken on a clear day. The terrain may require a lot of detail, but the sky is mostly featureless. A lossy compression process will in effect draw the terrain in detail and then say "paint the sky blue."

Mosaics can accomplish similar results without obfuscating the information being sent. The basic idea is to subdivide a large image into many smaller image tiles and then throw away the tiles that are unimportant to the image being conveyed. As an example, an image of the Earth from outer space can be converted into a mosaic of smaller images. The tiles that contain only black empty space don't need to be sent. That will reduce the amount of information needed to describe the Earth's surface in the same detail by about 20%.

This approach can be taken a step further by varying the resolution of each tile used to build an image. Regions of the image that have little detail can be represented by a low-resolution tile that has just one or a few pixels, while highly detailed regions can be represented with many high-resolution tiles. The receiver would need to work out how to assemble the tiles into a single image, but if the sender uses easily recognized reference images, this should be fairly straightforward to do.

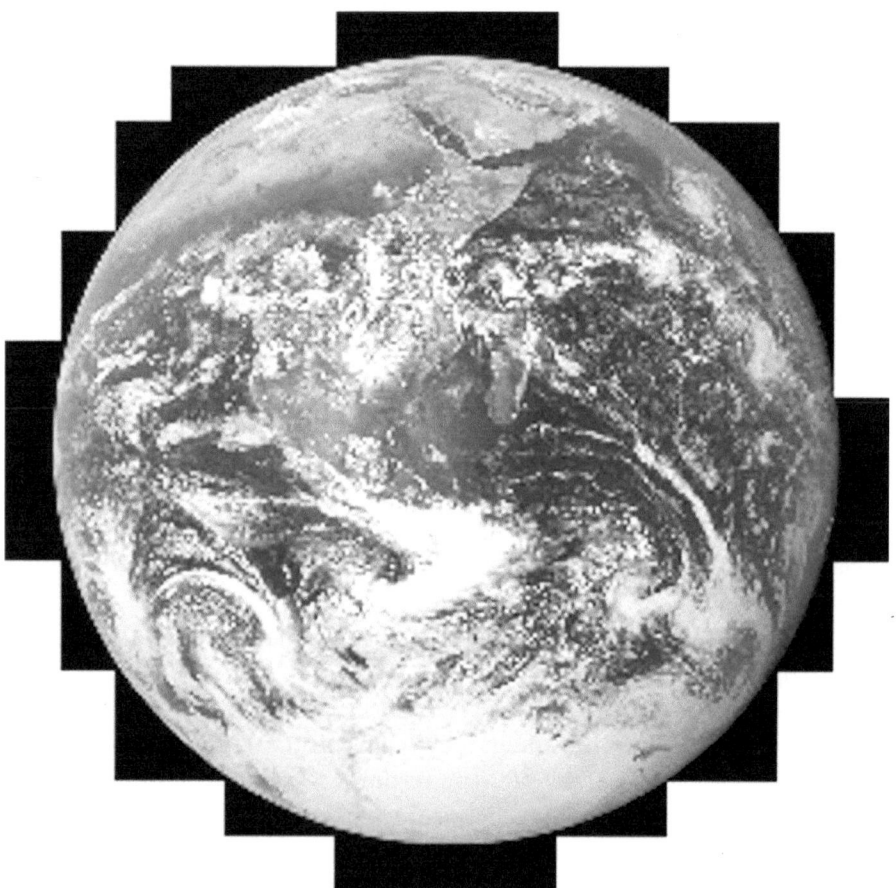

Fig. 10.17 The photo of the Earth taken by the Apollo 17 astronauts, displayed as a grid of 16 x 16 pixel tiles, with the tiles representing empty space omitted. (Image credit: Brian McConnell)

Color

What about color information? Adding color channels increases the amount of information needed to describe an image in proportion to the number of color channels used. This is complicated by the fact that the sender will probably have no idea what color bands the recipient of the images is sensitive to. The sender may want to send images with a large number of color bands, so that the receiver can map, or down-convert, color information to match the way their vision works.

One way to send lots of color information while minimizing the amount of information required to do so is to describe color information in lower resolution relative to the primary image. This is how color television worked prior to the introduction of digital/high-definition TV. The color TV system worked by describing color in relationship to a monochromatic signal for luminance (brightness). Let's explain how something like this could work with digital images.

One way to do this is to send a collection of images. One image is a high-resolution panchromatic (grayscale) image, while the rest are lower-resolution images (one for each color channel). Let's say that we want to include 16 color channels, each at 1/8th the size (1/64th the pixel count) of the high-resolution monochrome image. In this example, we can represent a hyper-spectral image using just 25% more information than we would need for a monochrome image. To explain how this works, see the following example.

The reconstructed image differs from the original only in that the color information is smeared a bit. If you zoom in to look at pixel level details, you'll notice that the red from the Golden Gate Bridge bleeds into surrounding pixels, whereas in the original boundaries between different colors are more precise. In this example, we are able to reduce the amount of information needed to describe a three-color-channel image compared to sending three full-resolution images by about 65%. The catch is that color information is encoded at a lower spatial resolution, so some information is lost.

Compression Algorithms

Algorithmic compression can yield much greater improvements, on the order of 10:1 compression ratios or more, but this requires the receiver to understand the algorithm being used to reverse the encoding process, which is at cross purposes with the primary design goal of making information easy to extract and comprehend. If the transmission also includes algorithms, it will

Fig. 10.18 A high-resolution RGB (three-color channel) image. (Image credit: Queen Mary 2 Golden Gate Crossing. Brian McConnell. San Francisco, CA (2007))

be possible to transmit programs that in turn assist the receiver in extracting highly compressed images (this is discussed in detail in the chapter Algorithmic Communication Systems).

Let's say that the sender has compressed many images using a lossless compression algorithm similar to the PNG format used on the web. They would include a program that implements the decompression process for this encoding. Once the recipient has learned to run the programs contained in the message, they would be able to use this program to run an arbitrarily complex decompression algorithm and use them to view compressed images.

There are reasons to be optimistic that images will be straightforward to extract if they are present in the transmission. The main reason is that the physics of photography and optics are universal. With a combination of automated and human analysis, it should be possible to find images relatively quickly if they are there. Understanding what the images contain, how they relate to each other, and what they mean may remain a mystery, but if the sender is motivated to make them easy to decode, we may be surprised how quickly these can be found.

Fig. 10.19 The color image is converted into a high-resolution monochrome (grayscale) image by averaging the red, green, and blue levels for each pixel. (Image credit: Brian McConnell)

Fig. 10.20 Color information is split out into lower-resolution images that encode the difference between the color channel value and the luminance value in the monochrome image (red-luminance, green-luminance, and blue-luminance). In this example, the color difference images are scaled down to 1/64th the detail of the original image. (Image credit: Brian McConnell)

How Many Images Can ET Send?

With this in mind, let's work out the number of images that could be sent via several different communication channels: a relatively slow radio or optical channel, a fast radio or optical channel, and inscribed matter. As we discussed earlier in the book, even a slow communication channel can deliver a surprisingly large amount of information over the course of a year.

Fig. 10.21 We can then reconstruct the original color image by recombining the downsampled color difference information with the high-resolution monochrome image. (Image credit: Brian McConnell)

For the purposes of this exercise, we'll use the following parameters to estimate the information carrying capacity of these circuits, measured in terms of one megapixel (million pixel) images. Note that there is nothing magic about a megapixel. Images may be sent in lower or higher resolution depending on what the sender is trying to convey. We'll use this as a baseline to give a sense of how many images a channel can carry.

Let's assume that redundant coding (forward error correction) and gaps in transmission coverage reduce transmission capacity by a factor of three and that images comprise roughly one third of the information sent. Divide total capacity by ten to get a rough idea of the information carrying capacity allocated to images when using these assumptions.

The slow communication channel operates at an average of 10,000 bits per second, comparable to an early dial up modem (31 gigabits/year, adjusted for the factors above).

The fast communication channel operates at an average of 10,000,000 bits per second, comparable to a modest home internet connection (31 terabits/year, adjusted for the factors above).

The inscribed matter system delivers an average of one gram of inscribed matter per year, with an information density of 10^{22} bits/kilogram or 10^{19} bits/gram (10^{18} bits/year, after adjusting for the loss factors above).

Next, let's assume that each pixel uses 10 bits of information to describe its luminance (brightness) level, which allows for 1024 discrete values. This is comparable to what we use for high-dynamic-range video content. You can apply your own assumptions about whether to use more or less bits per pixel, which will adjust the capacity of the system accordingly, but this is a good even number benchmark to start with.

Starting with monochromatic (grayscale) images, each megapixel image will require 10,000,000 bits of information (10 bits times 1 million pixels). We can work out the capacity of each channel as follows.

Slow channel: 3100 images per year
Fast channel: 3100,000 mages per year
Inscribed matter: 100,000,000,000 images per year

Next, let's account for color. To do that, simply divide these capacities by the number of color channels used for each image. If most images are sent in five color channels (such as near infrared, red/orange, yellow/green, blue, and ultraviolet), then divide the numbers above by five to adjust the number of images that can be sent. In reality, not all images are equal, so some will probably be sent in many color channels, others in monochrome. Also note that the sender could send color channel information in lower resolution, which will reduce the amount of extra information needed to describe color details (see the worked example above).

Then let's account for the effect of image compression. The numbers above are for uncompressed images. If the sender is using lossy compression for images, this will reduce the amount of information needed for each image by a factor of at least 5 to 10. As we discussed previously, lossy compression does result in the loss of information from the original image, but that may be okay for images where it is not necessary to provide an exact replica of the original. The sender can also control how much information is lost during the compression process. With less aggressive compression settings, the reproduced image is usually almost identical to the original.

One thing sticks out from this analysis, and that is that inscribed matter can deliver enormous amounts of information compared to radio or optical channels. Yet even if we are limited to slower communication channels, we could receive several thousand to several million images per year.

Checklist of Patterns to Watch for

Are uncompressed images evident when the data is viewed as a raw bitstream? (Planetary images should be relatively easy to spot if they are present.)

Do image collections/files appear to have regions of structured data or metadata, something that looks like a file header? (This could be used to convey information about image dimensions, encoding, color channels, set membership, etc.)

Do we see situations where the same image or collection is sent in different encodings, such as a calibration image that is sent in different resolutions, bit depths, etc.? If so, what varies between them? (This could hint at how metadata or header information is conveyed.)

Do images appear to be broken down into smaller subunits, such as tiles in a mosaic? If so, is there a system or nomenclature to assist in reassembling them into larger collections?

Do images appear to have spectral information attached to or associated with them, such as diagrams of Fraunhofer lines?

Do there appear to be sets of images that are slightly different? Sets of images like this may be used to represent color channels.

Do we see collections of data that could be compressed image data?

Planetary Images

Saturn as Seen by Cassini

Fig. 10.22 Saturn's north pole and ring system, as seen by the Cassini probe. Original color image. (Image Credit & License: NASA/JPL/SSI; Composition: Gordan Ugarkovic)

Fig. 10.23 Saturn as seen by the Cassini probe, dithered grayscale encoding where each pixel's luminance value is represented by a 16 x 16 grid of binary digits

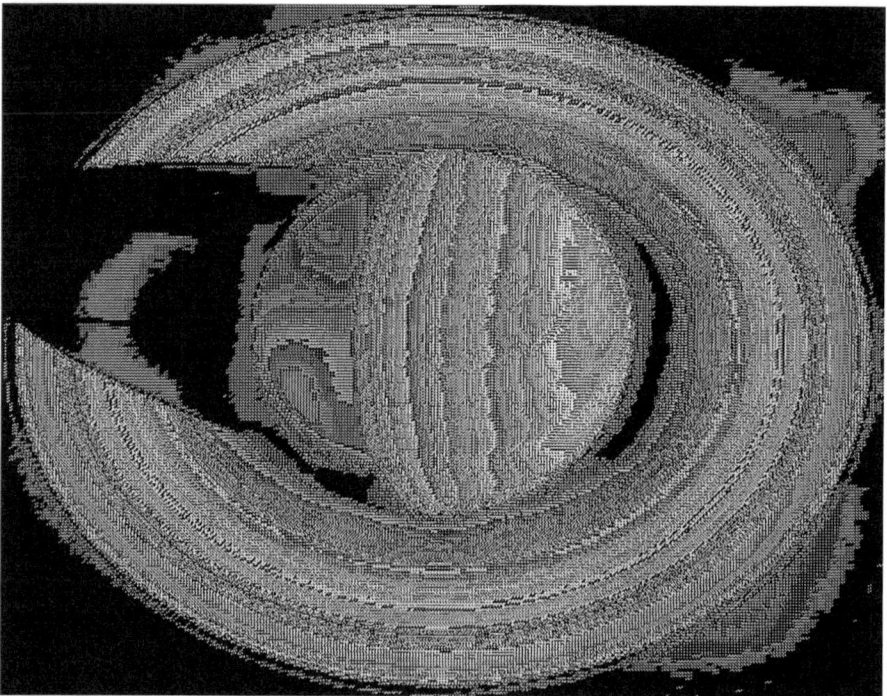

Fig. 10.24 Saturn as seen by the Cassini probe, this time in a 3 x 3 grayscale encoding, where each pixel's luminance value is represented by a 3 x 3 grid of binary digits. Note that major features of the image are visible and that the aspect ratio of the image is not distorted

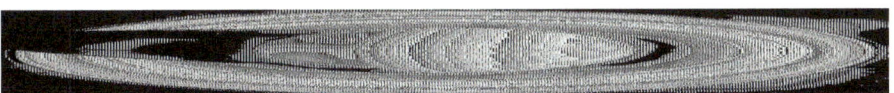

Fig. 10.25 Saturn as seen by the Cassini probe, in an 8-bit linear grayscale encoding. Notice that the major features of the image are readily visible, but are stretched out by an 8:1 ratio, since each pixel is represented by eight binary digits

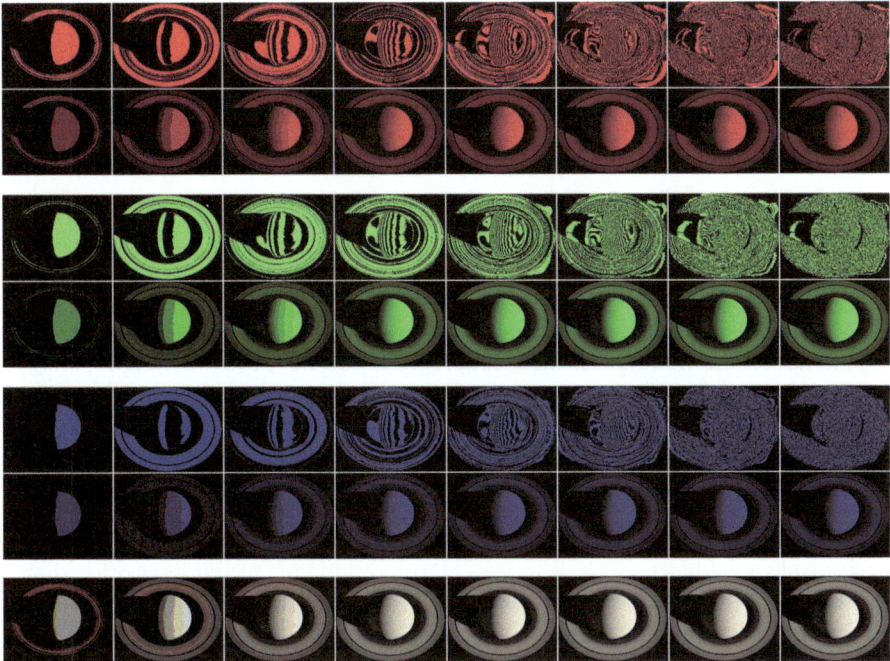

Fig. 10.26 Saturn as seen by the Cassini probe, in a three-channel (RGB) bitplane representation, proceeding from the most significant bit on the left to the least significant bit on the right. Notice that major features of the image are visible even with just the least significant bit and that luminance and color detail improve as more bitplanes are added in

A Jupiter Vista from Juno

Fig. 10.27 A Jupiter Vista from Juno. Original color image. (Image Credit: NASA/JPL-Caltech/SwRI/MSSS; Processing & License: Kevin M. Gill)

Fig. 10.28 A Jupiter Vista from Juno. (Image Credit: NASA/JPL-Caltech/SwRI/MSSS; Processing & License: Kevin M. Gill. Shown in a grayscale, dithered encoding where each pixel's luminance value is represented by a 16 x 16 grid of binary digits. While this is an inefficient encoding and requires 32 times as much information as a linear 8-bit encoding, it is essentially a self-decoding representation)

Fig. 10.29 A Jupiter Vista from Juno. (Image Credit: NASA/JPL-Caltech/SwRI/MSSS; Processing & License: Kevin M. Gill. Shown in an 8-bit linear grayscale encoding, where each pixel's luminance value is represented by eight binary digits. Notice that the major features, such as the curvature of Jupiter, are visible, although fine luminance detail is not obvious)

Fig. 10.30 A Jupiter Vista from Juno. (Image Credit: NASA/JPL-Caltech/SwRI/MSSS; Processing & License: Kevin M. Gill. Shown in a three-channel (RGB) bitplane representation, proceeding from the most significant bit on the left to the least significant bit on the right. A recombined color image is shown in the bottom row. Notice that major features of the image are visible in the most significant bitplanes and that luminance and color detail improve as more bitplanes are added in)

Landscapes

Green Bank Telescope

Fig. 10.31   Green Bank telescope, Green Bank WV (2016). Original image. (Image credit: Brian McConnell)

Fig. 10.32 Green Bank telescope, dithered grayscale encoding. (Source image credit: Brian McConnell)

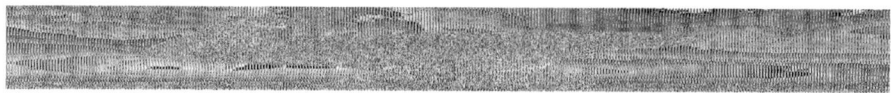

Fig. 10.33 Green Bank telescope, linear 8-bit grayscale encoding. (Source image credit: Brian McConnell)

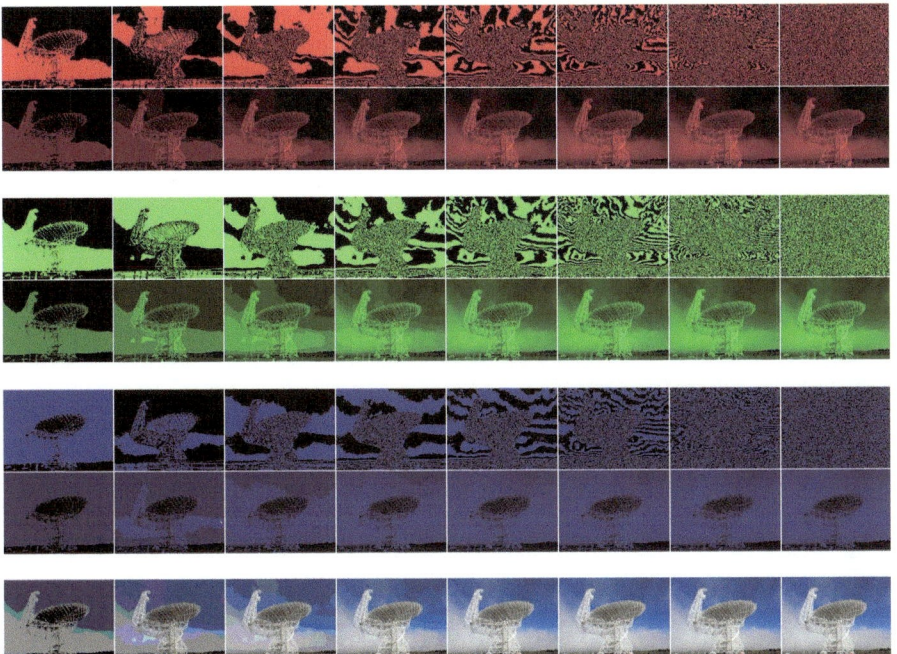

Fig. 10.34 Green Bank telescope, three-channel (RGB) bitplane encoding, proceeding from the most significant bit on the left to the least significant bit on the right. A recombined color image is shown in the bottom row. Notice that major features are readily visible with just the least significant bit and that fine luminance and color detail improve as less significant bits are included

The Great Pyramid of Giza

Fig. 10.35 Photograph of the Great Pyramid of Giza, original color image. (Image credit: Nina Aldin Thune via Wikimedia Commons / Creative Commons License)

Fig. 10.36 Photograph of the Great Pyramid of Giza, Image credit: Nina Aldin Thune via Wikimedia Commons / Creative Commons License. Shown in a grayscale dithered encoding where each pixel's luminance value is represented by a 16 x 16 grid of binary digits

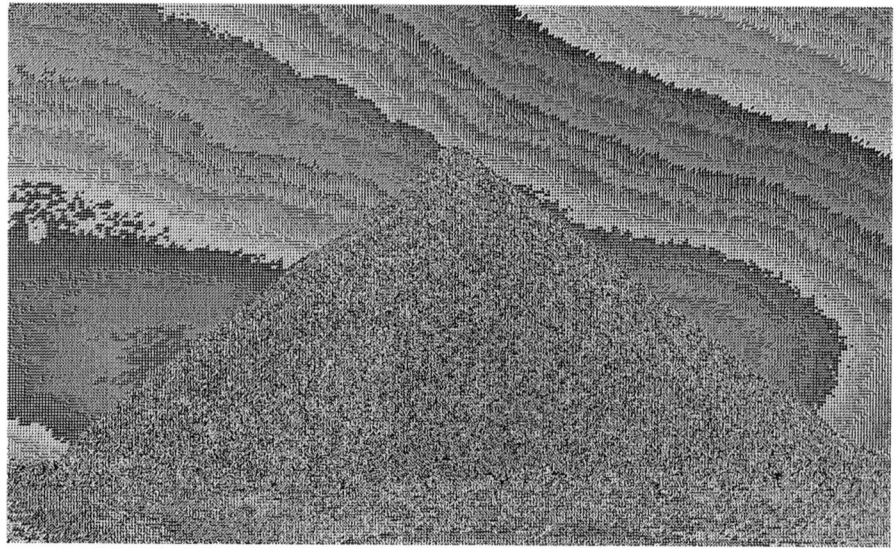

Fig. 10.37 Photograph of the Great Pyramid of Giza, Image credit: Nina Aldin Thune via Wikimedia Commons / Creative Commons License. Shown in a 3 x 3 grayscale encoding where each pixel's luminance value is represented by a 3 x 3 grid of binary digits. Note that the pyramid is visible, but luminance detail is distorted

Fig. 10.38 Photograph of the Great Pyramid of Giza, Image credit: Nina Aldin Thune via Wikimedia Commons / Creative Commons License, shown in a three-channel (RGB) bitplane representation, proceeding from the most significant bit on the left to the least significant bit on the right. A recombined color image is shown on the bottom row. Notice that the major features of the image are visible even with only the most significant bit and that luminance and color detail improve as more bitplanes are added in

Life Forms

Praying Mantis

Fig. 10.39 Praying mantis (2005). (Image credit: Shiva shankar. Taken at Karkala, Karnataka, India. Creative commons via Wikimedia. Original color image)

Fig. 10.40 Praying mantis (2005). (Image credit: Shiva shankar. Taken at Karkala, Karnataka, India. Creative commons via Wikimedia. Dithered grayscale encoding where each pixel's luminance value is represented by a 16 x 16 grid of binary digits)

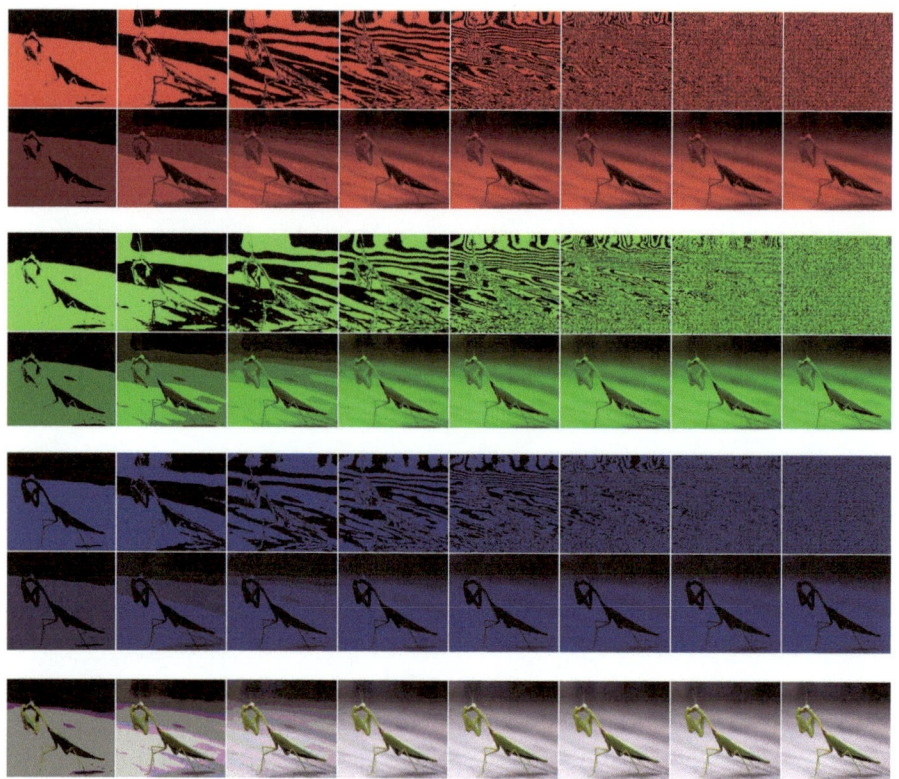

Fig. 10.41 Praying mantis (2005). (Image credit: Shiva shankar. Taken at Karkala, Karnataka, India. Creative commons via Wikimedia. Three-channel (RGB) bitplane representation, proceeding from the most significant bit on the left to the least significant bit on the right. A recombined color image is shown in the bottom row. Notice that major features of the image are visible with just the most significant digit and that luminance and color detail improve as more bitplanes are added)

Purple Striped Jellyfish

Fig. 10.42 Purple striped jellyfish. Fred Hsu. (Photo taken and uploaded by user) Sea Nettle Jelly, Jellyfish, Monterey Bay Aquarium, California. Original color image.

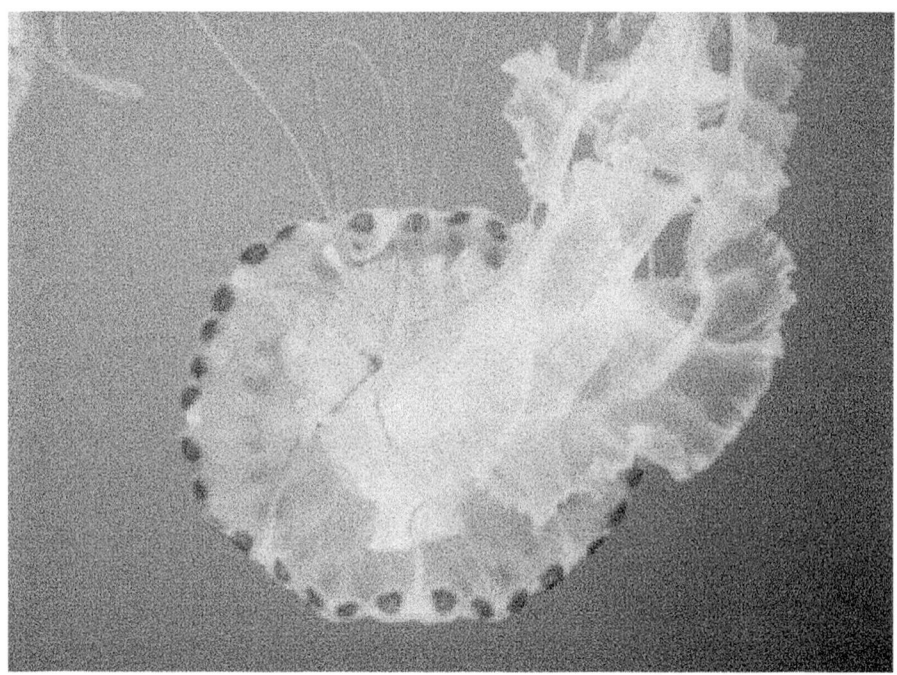

Fig. 10.43 Purple striped jellyfish. Fred Hsu. (Photo taken and uploaded by user) Sea Nettle Jelly, Jellyfish, Monterey Bay Aquarium, California. Grayscale, dithered encoding where each pixel's luminance value is represented by a 16 x 16 grid of binary digits.

Fig. 10.44 Purple striped jellyfish. Fred Hsu. (Photo taken and uploaded by user) Sea Nettle Jelly, Jellyfish, Monterey Bay Aquarium, California. 8-bit linear grayscale encoding, where each pixel's luminance value is represented by eight binary digits. Notice that details in the image are obfuscated, but vertical striping hints at the number of bits used to represent each pixel.

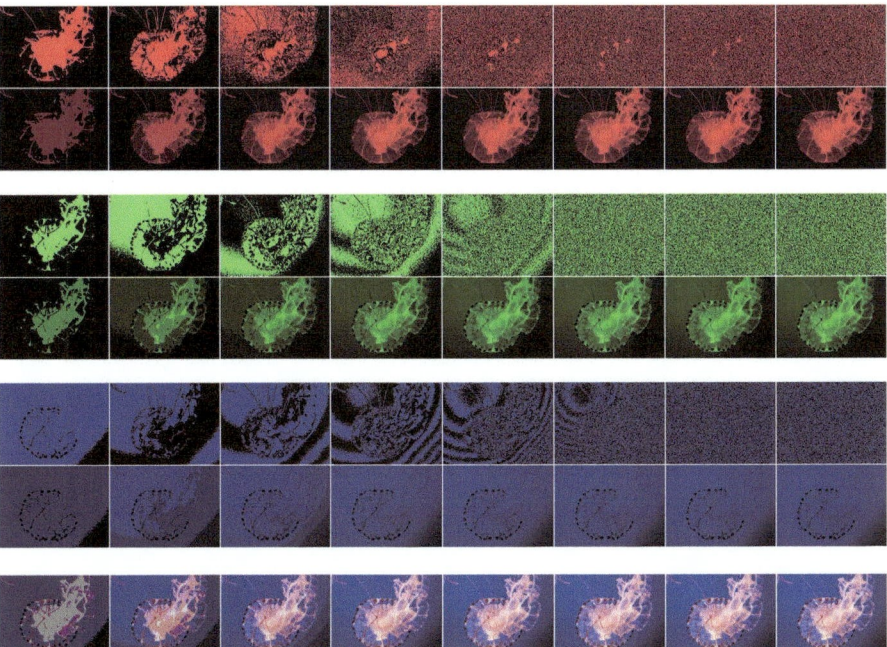

Fig. 10.45 Purple striped jellyfish. Fred Hsu. (Photo taken and uploaded by user) Sea Nettle Jelly, Jellyfish, Monterey Bay Aquarium, California. Three-channel (RGB) bit-plane representation, proceeding from the most significant bit on the left to the least significant bit on the right. A recombined color image is shown in the bottom row. Notice that luminance and color information become more detailed as more bitplanes are added in from left to right.

Visual Art

Edward Hopper "Groundswell" (1939)

One of the useful features of bitmaps is that once the recipient has figured out how to decode them, they can be used to represent not just photographs, but any visual representation or image. Here we examine artist Edward Hopper's painting "Groundswell," as shown in a variety of encodings.

Fig. 10.46   Edward Hopper "Groundswell" (1939). (Original image)

Fig. 10.47 Edward Hopper "Groundswell" (1939), shown in a dithered grayscale encoding, where each pixel's luminance value is represented by a 16 x 16 grid of randomly placed binary digits. This encoding while it is inefficient is essentially self-decoding

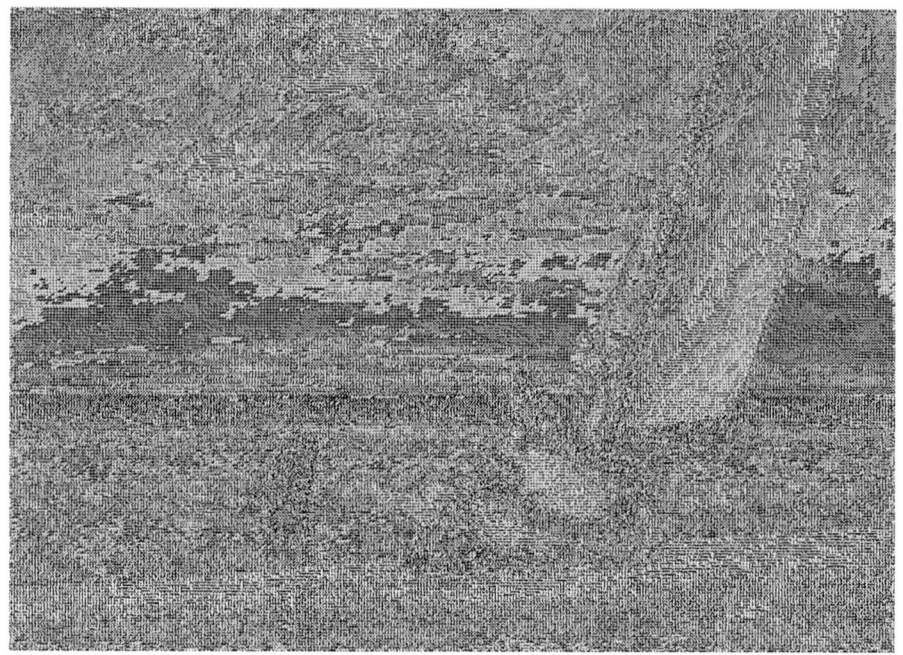

Fig. 10.48 Edward Hopper "Groundswell" (1939), shown in a 3 x 3 grayscale bitstream encoding, where each pixel's luminance value is encoded as a 3 x 3 array of binary digits. Notice that while some large features, such as the boat's sail, are visible, the overall content of the image is not immediately obvious

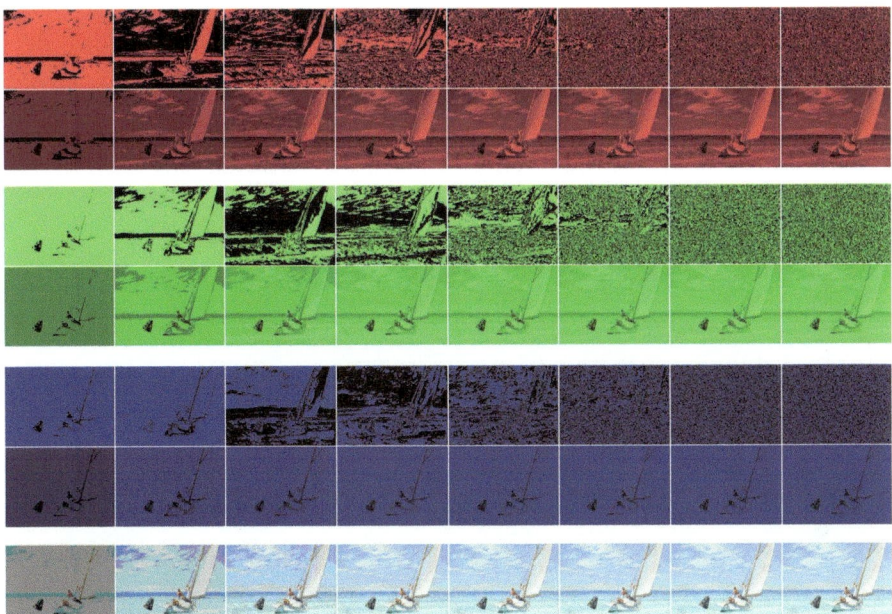

Fig. 10.49   Edward Hopper "Groundswell" (1939). Displayed in a three-channel (RGB) bitplane representation, preceding from the most significant bit on the left to the least significant bit on the right. A recombined color image is shown in the bottom row. Notice how, as more bitplanes are added, the luminance and color detail improve from left to right

Visual Art

The Tower of Babel (1563)

Fig. 10.50 Pieter Bruegel the Elder, "The Tower of Babel" (Vienna). 1563. Color scan courtesy of the Google Art Project

Fig. 10.51 Pieter Bruegel the Elder, "The Tower of Babel" (Vienna). 1563. Dithered grayscale encoding where each pixel's luminance value is represented by a 16 x 16 grid of binary digits

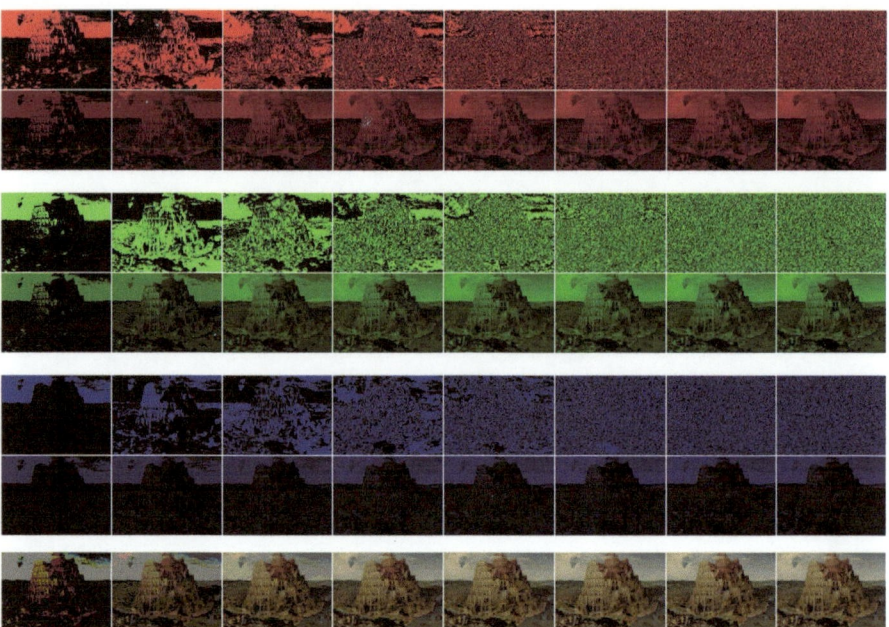

Fig. 10.52 Pieter Bruegel the Elder, "The Tower of Babel" (Vienna). 1563. Three-channel (RGB) bitplane representation, proceeding from most significant digit on the left to least significant digit on the right. A recombined color image is shown in the bottom row. Notice that major features of the image are visible with just the most significant digit and that luminance and color detail improve as more bitplanes are added in

11

Three-Dimensional Images and Models

Just as additional color bands can be added by sending one image for each color band, there are a number of ways to transmit three-dimensional imagery. This will enable a sender to describe complex three-dimensional scenes, such as holographic models of organisms and landscapes.

Stereoscopic Imagery

The *parallax method*, or *stereoscopic imagery*, is one way to do this. This method works by taking photographs of a target from two or more different positions. Objects in the foreground will shift relative to background objects, and the degree to which they shift provides distance or depth information. The parallax method for measuring distance should be familiar to an astronomically literate civilization and therefore is a pattern we should look for.

Stereoscopic imaging will be useful for communicating information about the environments and objects being photographed and, in some cases, might be critical to communicating their structure. We can usually infer depth and dimensions from two-dimensional scenes due to our familiarity with the items within them, something that would not necessarily be true for photographs of an alien environment or organism. This method can be further extended to three-dimensional models by sending stereoscopic images taken from multiple directions around the object being photographed.

If we were composing a message to be transmitted to other civilizations and wanted to include a catalog of organisms on Earth, this would be an excellent way to communicate their shape along with their appearance in a flat image.

© The Author(s), under exclusive license to Springer Nature Switzerland AG 2021
B. S. McConnell, *The Alien Communication Handbook*, Astronomers' Universe,
https://doi.org/10.1007/978-3-030-74845-6_11

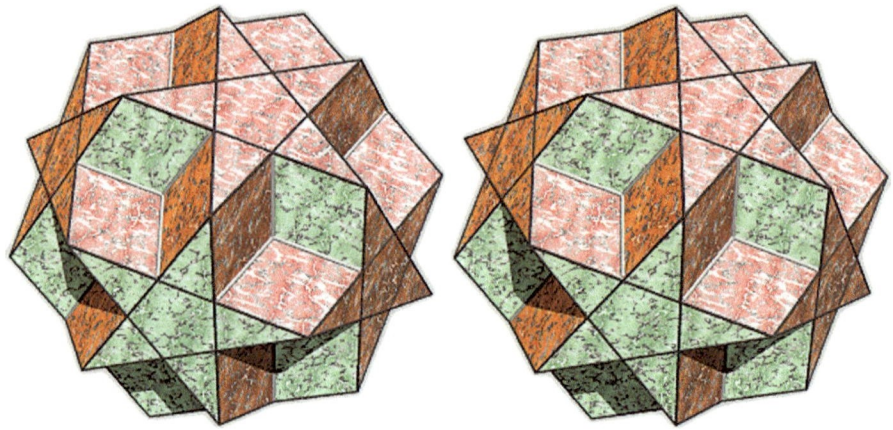

Fig. 11.1 An example of a stereoscopic image using the "eyes crossed" method. If you cross your eyes until the two images are fused, a three-dimensional view of the object will emerge. (Image courtesy Mark Newbold (Stereoscopic image, Mark Newbold, https://dogfeathers.com/3d/lookcrossed.html))

The information cost of doing so increases by 12, with 6 pairs of stereoscopic images of the subject, from 6 orthogonal directions if the goal is to send enough data to construct a crude holographic model.

There are several other techniques that can be used to describe three-dimensional systems, but this approach has the advantage that it's just a larger set of images. If the receiver doesn't figure out that they are stereoscopic images, they can still view them as sets of unpaired images taken from different perspectives.

Point Clouds

What if we want to describe a complex shape or one that has hidden structures beneath its surface? To do this, we will need to send a three-dimensional model of the object. *Point clouds* are one way to do this without using a complex modeling language. To describe a point cloud, all one needs to do is to send a list of three-dimensional coordinates.

This is an economical way to depict complex three-dimensional structures, especially if the shape described is a surface. Let's say that the sender wants to send a point cloud that describes the skeleton of an organism. The hypothetical point cloud maps to a volumetric display that measures 4096 (4K) units on a side. This would require 36 bits (12 bits per axis) to describe each point coordinate. To send a reasonably detailed point cloud with between 100,000 and 1,000,000 points would require 3,600,000–36,000,000 bits of information (0.45–4.5 megabytes), similar to the amount of data needed for a high-resolution photograph. The number of points described would depend on what level of detail is required to reproduce the object with decent fidelity. The data would be sent in a form like:

$$\big(\big((x_1)(y_1)(z_1)\big)\big((x_2)(y_2)(z_2)\big)\big((x_3)(y_3)(z_3)\big)....\big((x_n)(y_n)(z_n)\big)\big)$$

What could the sender do to assist the receiver in understanding how to read point clouds? One thing the sender could do is to send reference models that describe basic geometric shapes, such as spheres, cylinders, and so on. While it would take some guesswork to understand what this data represents, the basic encoding scheme, as with bitmaps, is simple. Once understood, the recipient would be able to render arbitrarily complex shapes.

Point clouds can be extended to include luminance and color information. To do this, one would add additional numbers to describe the relative intensity of each color channel for a given point. With this approach, one can create arbitrarily detailed 3D models that also include brightness and color

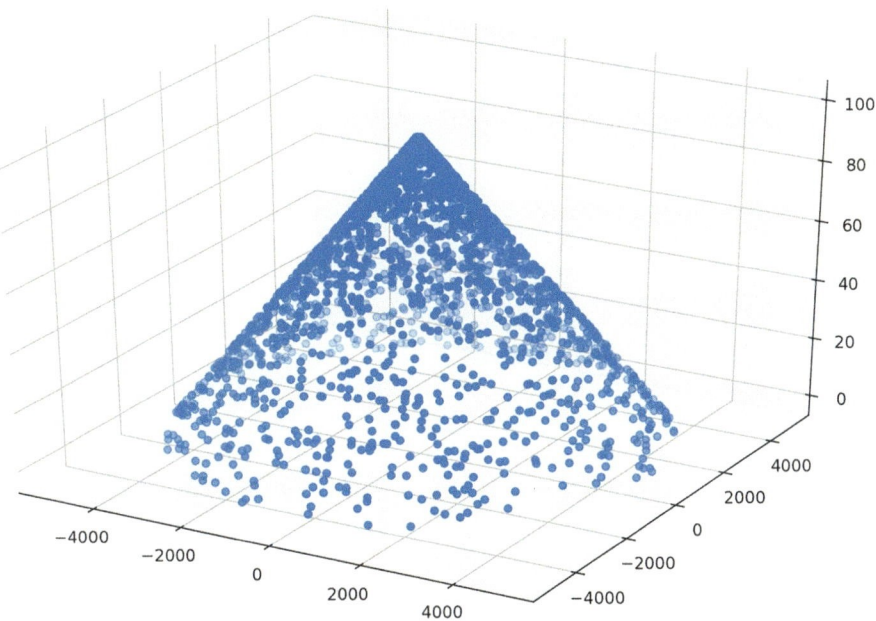

Fig. 11.2 A point cloud depicting a cone. (Image credit: Brian McConnell)

information. The encoding scheme remains simple, as the point cloud simply consists of a list of numbers. A three-color channel point cloud, for each point, would contain three numbers for x, y, and z coordinates and three for relative intensities for the three color channels. This increases the information footprint, but the sender can control what level of detail and how many points are used to describe an object.

In the example above, we use a Cartesian (three-dimensional) coordinate system. Point clouds can also be described in other coordinate systems, such as spherical and cylindrical coordinates. A Cartesian coordinate system is attractive because it can be described using only integer numbers. On the other hand, a spherical coordinate system will be useful for describing spherical objects, such as a 3D model of a planet's surface.

Describing A 3D Model of the Earth

Let's suppose we are composing a message and want to send a 3D model of the earth with the following attributes:

- 5 million data points, 1 per 100 square kilometers (10 kilometers by 10 kilometers grid) on average, though some regions can have a higher sample density than featureless areas like the ocean.
- Radius, accurate to 10 centimeters or less. A 32-bit number will allow 4 billion unique values, enough to allow millimeter-level precision at this scale.
- 24 bits to represent each angular coordinate, which allows for 2-meter-pixel size at the equator.
- 5 color channels (near infrared, red, green, blue, ultraviolet), 8 bits per color channel.

This works out to 2 gigabits, or 250 megabytes, to represent a 3D model of the Earth's surface. At a 10-km-per-pixel resolution, it is possible to map the Earth's surface in reasonable detail. The points don't need to be uniformly distributed either, as the model could use lower point densities over featureless areas like the ocean while using higher point densities to map out important sites such as urban areas, mountainous terrain, etc.

Polygon Meshes

Point clouds are also a simple way to send three-dimensional data, but one limitation they have is they do not define how the points are connected to each other. The receiver needs to infer this, which can be a problem with complex structures, especially multilayered structures, as it may not be obvious which points are connected to which.

An arbitrarily complex 3D surface can also be described as a mesh of triangles. Each triangle can be defined by a set of three coordinates in three-dimensional space. This would take the form of a nested data structure as follows:

$$\left((t_1)(t_2)(t_3)... \right)$$

where t_n is described as a triplet of three-dimensional coordinates, with a three-dimensional coordinate for each vertex of the triangle, as shown below:

$$\left(\left((x_1)(y_1)(z_1) \right)\left((x_2)(y_2)(z_2) \right)(x_3)(y_3)(z_3) \right)$$

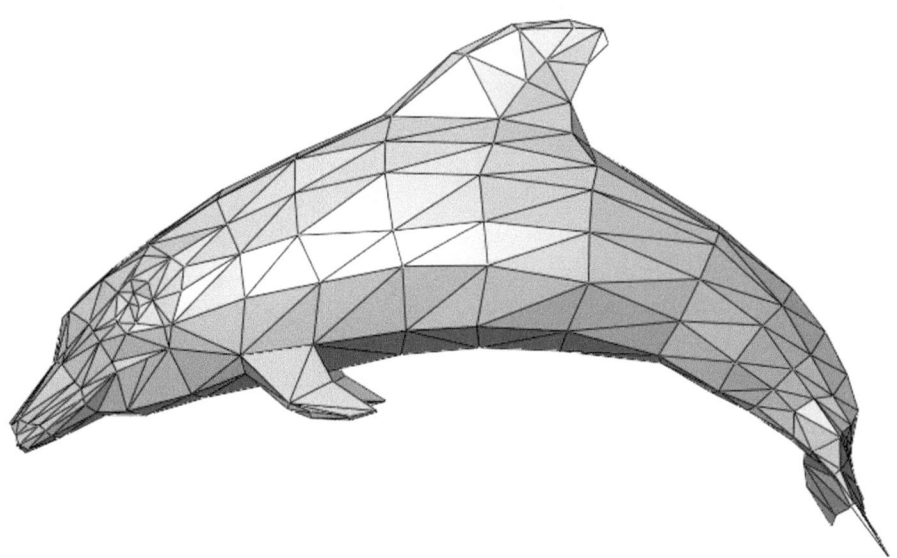

Fig. 11.3 An example of a low-resolution polygon mesh depicting a dolphin as a three-dimensional model. (Image credit: Wikimedia Commons)

This notation is more verbose, as a given coordinate may be referenced several times if it is used in multiple triangles, but this approach eliminates ambiguity about how nodes in the surface are interconnected. This format requires roughly three times as much information as a point cloud but eliminates ambiguity about surface boundaries.

Like point clouds, the sampling density can be adjusted to achieve the desired level of fidelity in describing a three-dimensional structure and can be varied as needed for different parts of the object being depicted. A simple geometric object can be described with just a few triangles (a pyramid can be described with six triangles, one for each face, and two for the base). Complex organic structures or natural scenes such as a landscape may require a dense mesh to describe accurately.

Unlike point clouds, polygon meshes can be used to describe complex structures that are nested within each other. Let's say the sender wants to describe organisms from their world. While exterior photos are useful, they don't reveal much about an organism's anatomy, unless it is translucent like a jellyfish. A sender could use a 3D model like this to describe an organism's internal structures such as its skeleton (if it has one), internal organs, and other details that would be invisible in an external photo.

One disadvantage of polygon meshes is they are not optimal for describing curved surfaces without using a large number of polygons. The receiver can apply curve-fitting algorithms to low-resolution meshes to partially correct for this, but this involves making guesses about the geometry of the depicted object, which may not always be correct.

Programmatically Generated Objects

If the data stream includes algorithms (see the chapter Algorithmic Communication), the sender might create programs that generate dense mesh networks as output to be rendered in a 3D display system. This will be especially useful in generating shapes that can be defined tersely in mathematical form, as well as describing complex objects that can be assembled from a collection of simpler primitive objects.

Take the example of the cone described as a point cloud. The cone is represented in a point cloud by a large number of three-dimensional coordinates, when it could instead be represented by just three pieces of information: the radius of the cone's base, the coordinates of the center of the base, and the coordinates of the tip of the cone. A compact computer program could accept these inputs and generate an arbitrarily dense point cloud or mesh as output.

So instead of wasting bandwidth sending this shape as a point cloud, the sender would reference a function that generates a high-resolution mesh, which in turn can be rendered as a 3D scene.

The use of algorithmically generated three-dimensional objects will also enable the sender to describe complex scenes that are composed of many reusable objects, such as a large settlement or structure, while minimizing the amount of information needed to do so. The sender could also use algorithms to upscale low-resolution meshes, for example, to fit curves to a simplified model, such as the dolphin shown earlier, and then generate a high-resolution mesh as output. Algorithmically generated 3D models could also be fed into an input/output interface (see Chap. 9 on Algorithmic Communication Systems) and, if so, could interact with the user.

Checklist of Patterns to Watch for

Do there appear to be many images of an object that are taken from different perspectives? If so, try various approaches to treat these as stereoscopic images.

Are there sets of stereoscopic images that are taken in orthogonal directions around an object? If so, try to generate a 3D model from this data.

Do there appear to be sets of three or more values in a structured data set? This could be data for a point cloud.

Do there appear to be sets of polygons with shared vertices? This could be a sign of polygon mesh.

Are there algorithms in the transmission that appear to generate point clouds or other multi-dimensional data sets as output?

Are there algorithms that write data for point clouds or meshes to a specific set of memory addresses or variables? If so, this might hint at the use of a reserved block of memory for use as a 3D display interface (see the discussion about input/put interfaces in Algorithmic Communication Systems).

12

Four Dimensions (Video and Simulations)

So far we have described various ways that a sender could depict two-dimensional and three-dimensional scenes using images and simplified mathematical models. These are adequate for representing still scenes, but what if the sender wants to describe a scene that changes over time?

Motion Pictures

An easy way to do this is to transmit a series of images, in the same way that motion pictures work. The advantage here is that the recipient can view the content as a series of images, even if they do not immediately recognize that it is a motion picture. The direction of time can also be made clear by including motion pictures, such as smoke rising from a fire, that only make sense when played in one direction.

The main disadvantage of this format is that it requires a lot of information to describe a complex scene over time. Let's say that the sender wants to include a short video of an important scene or event. In this example, they use the following parameters:

- Moderate image resolution – 1000 x 1000 pixels, 8 bits per pixel, about 50% of the resolution for standard high definition TV
- Four color bands – near infrared, red/orange, yellow/green, blue/violet
- Duration – 60 seconds
- Frame rate – 10 per second

© The Author(s), under exclusive license to Springer Nature Switzerland AG 2021
B. S. McConnell, *The Alien Communication Handbook*, Astronomers' Universe,
https://doi.org/10.1007/978-3-030-74845-6_12

This scene will require 600 still images, each of which will require 4 megabytes (32 megabits) of information, for a total of 2.4 gigabytes (19.2 gigabits) of information. This is significantly more information than is required for a still image – and this is a pretty conservative example. As the resolution, frame rate, and duration increase, the information requirements increase dramatically. Because this type of content requires much more information to represent, the sender will probably be forced to make tradeoffs about what to send. If the same scene or process can be represented with a small set of still images, is it worth it to send a higher-fidelity motion picture at the expense of crowding out other things they would like to include?

How Much Data Is Needed to Transmit a Movie?

In keeping with the theme of contact with another civilization, let's calculate how much information is needed to transmit the sci-fi classic *2001: A Space Odyssey*.

The film has a running length of 7440 seconds. Most films at that time ran at a rate of 24 frames per second, so that works out to 178,560 frames. If each frame is recorded at the resolution of a high-definition TV broadcast (about two million pixels), that works out to 357,120,000,000 pixels. At 24 bits per pixel (3 color channels, 8 bits per pixel), that works out to about 8.6 trillion bits (terabits) of data.

Next let's estimate how much information is needed to represent the audio content of the film. For stereo audio, recorded at a 40,000 Hz sample rate with 16 bits resolution per sample, that works out to 9.52 billion bits (gigabits) of data, roughly 1/1000th the information footprint of the video content.

We can then estimate how long it will take to transmit, or how much inscribed matter is needed to represent it. Let's compare three options, a slow radio/optical channel that transmits data at 1 megabit per second in total, a fast radio/optical channel that transmits at 1 gigabit per second, and inscribed matter encoded at 10^{22} bits per kilogram.

The slow channel will be able to transmit the film in about 99 days, while the fast channel will take just 2.38 hours. If inscribed matter is used to deliver the information, less than one millionth of a gram will be needed, meaning that a million movies could be fit into a mass weighing less than a 1/30th of an ounce. This shows that inscribed matter has clear advantages over electromagnetic transmissions, especially for the delivery of information that is more archival in nature.

Time to transmit at 1 megabit/sec: 99.2 days.

Time to transmit at 1 gigabit/sec: 2.38 hours.

Inscribed matter at 10^{22} bits/kg: ~ 1 microgram (1 millionth of a gram).

Note that these estimates assume that the video data is not compressed, so this is a conservative estimate of how much information will be required to represent a typical film or video sequence.

Video Compression

Because motion pictures require so much information to represent, there is an incentive to look for ways to reduce this footprint. There are a couple of avenues the sender can explore. One is an implicit approach to data reduction, and the other is to use compression algorithms, in the case of a message that is composed in part with computer programs.

Implicit Compression

Here we want to reduce the amount of information needed to represent a given sequence of images, while enabling the receiver to reconstruct the original or an approximation of it. Most motion picture sequences contain a lot of redundant information, as each frame changes incrementally from the previous one. One way the sender can reduce the amount of information required is to transmit key frames in high resolution and to send intermediate frames that don't change very much in lower resolution. Let's revisit the example shown above.

Instead of sending each 1000 x 1000 frame in full resolution, we send only *key frames* (frames that are significantly different from the previous frame) in full resolution. Intermediate frames are sent at 100 x 100 resolution. The receiver is left to develop an upscaling algorithm, using information from the key frame and downsampled intermediate frames to reconstruct the full-resolution intermediate frames. Key frames are sent on average once every 20 frames. The compression ratio resulting from this process can be calculated with the equation below:

$$C = \frac{(n_k + n_i) \times pixels_k}{(n_i \times pixels_i) + (n_k \times pixels_k)}$$

This approach will reduce the information footprint of the uncompressed 1-minute image sequence by a factor of 16.8–1 for this example. If the receiver is unable to devise an upscaling algorithm to reconstruct the intermediate frames, the downsampled frames will be plainly visible, so in the worst case, they would be able to view the sequence with degraded resolution for the intermediate frames.

The sender can also decide how aggressively to compress the image sequence and can do so by adjusting the ratio of downsampled frames to key frames, as well as the downsized frame resolution. These parameters can be adjusted throughout the image sequence to optimize for maximum compression and quality based on image content. More aggressive compression will generally result in more loss of information, which is a typical tradeoff with lossy compression schemes.

Algorithmic Compression

The previous example shows that uncompressed motion pictures will require large amounts of information to represent. This is another example where it will be helpful to include computer programs in the message, in this case to implement video compression algorithms to reduce the amount of information needed.

Video compression algorithms take advantage of the fact that most information in a sequence of images remains the same and encodes the differences between frames to minimize the amount of information needed to reconstruct the original sequence of images. This is not as simple as just subtracting one frame from another, so the algorithms used to do this can be quite complex, and not the sort of thing that can be calculated by hand or explained in shorthand form. One of the reasons computer programs are so useful is they can mindlessly compute an arbitrarily complex sequence of instructions and repeat them billions of times.

Note also that if the message contains algorithms, the sender could also choose to include an upscaling algorithm that is designed to work with downsampled image sequences as described in the previous section. This would give the receiver two paths to accessing higher-resolution video content, either by developing their own upscaling algorithm or by using the upscaling algorithm defined by the transmitter.

Simulations

Other types of scenes or processes can be depicted using computer simulations. This is another example of a situation where an algorithmic communication system will be useful.

Let's say that the sender wants to send a detailed gravitational model of their solar system. This can be done using a numerical simulation. They would send the masses, starting positions, and starting velocities of every object in the model. This requires seven pieces of information for each object: three for its spatial location of its starting point in three-dimensional space, three for its velocity vector in three-dimensional space at the start of the simulation, and one for the mass of the object. The simulation calculates the gravitational interaction between every object and then from that calculates each object's position and velocity at the next time interval in the simulation. It then repeats this process ad infinitum.

To start the simulation, the program would load the initial conditions for each object. For the purposes of this example, we assume that each of the seven coordinates uses a 64-bit number (8 bytes) to represent its mass, position, and velocity with a high degree of accuracy at the distance scales needed for this type of simulation. Each object in the simulation then requires 448 bits (56 bytes) of information to describe its initial condition.

Let's say that we wanted to create such a model for our solar system and include all of the planets, dwarf planets, moons, major asteroids, and comets. This catalog would include hundreds of objects, and the resulting simulation would model their orbits with great precision. Despite that, the amount of information required is modest. A simulation that modeled the orbits of 1000 objects would require only 56 kilobytes of information to initialize, not including the instructions for the simulation program, which will have a similarly small footprint.

We can see then that simulations will be an economical way to describe three-dimensional systems that evolve over time, especially systems that are composed of a large number of particles or agents that interact with each other according to simple rules. Simulations also enable the user to pose "what if" questions, where they can probe the behavior of the simulation by modifying the parameters of the objects being modeled.

Numerical simulations do not produce a perfect replica of the system being modeled, as there is usually some error in the starting conditions. Simplifying assumptions, such as ignoring the effects of radiation pressure or omitting less massive objects, will also result in errors that accumulate the longer the simulation runs. That said, they are powerful tools for understanding complex systems and can produce accurate predictions for long periods of time.

13

Sound

Sound is the variation of pressure in an atmosphere or fluid over short time scales. Hearing has evolved in most animal species on Earth, and in the case of marine animals, it is a dominant sense. So it would not be surprising to discover that sound is an important component of an extraterrestrial message. Even if it is not important to how an ET species communicates, sound is an interesting physical phenomenon and something that could still be easily discovered via instrumentation and measurement.

As discussed previously, sound is the result of pressure in a medium changing over short (sub-second) time scales. One way to record sound is simply to plot pressure versus time. The way we typically do so is by graphing pressure in the vertical axis and time in the horizontal axis. This is simplifying things a bit, but that's the general idea. The result is a plot like.

The advantage of this approach is the receiver only needs to guess two parameters to successfully decode the audio, the *sample rate* (the number of times the pressure is measured, or sampled, in a given time interval), and the *sample width* (the number of bits used to represent each sample).

Humans can typically hear sounds ranging from 20 to 20,000 Hertz (one Hertz equals one cycle per second). To accurately record audio in this range, the audio needs to be sampled at twice the upper frequency limit, or 40,000 times per second. Compact discs and other digital audio formats typically have a 44,000 Hertz sample rate for this reason. This range is specific to humans. Many animals can hear sounds at much higher frequencies, and some can use sound to image their environment and navigate via echolocation. Because of this, it is a bad idea to bake assumptions related to human hearing range into an encoding scheme for audio.

© The Author(s), under exclusive license to Springer Nature Switzerland AG 2021
B. S. McConnell, *The Alien Communication Handbook*, Astronomers' Universe,
https://doi.org/10.1007/978-3-030-74845-6_13

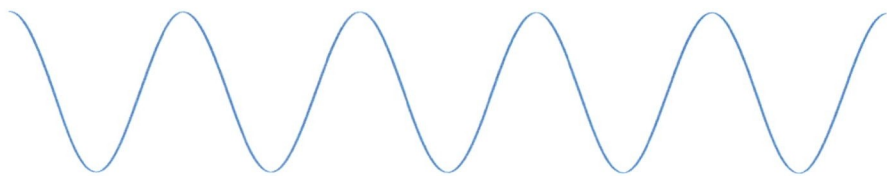

Fig. 13.1 A plot of a sine wave. The x-axis denotes time, while the y-axis denotes amplitude (the strength of the signal). (Image credit: Brian McConnell)

Table 13.1 Information required for different types of audio samples

Sample rate	Sample size	One-minute sample	Remarks
8000 Hz	8 bits	3.84 megabits	Telephone audio
20,000 Hz	16 bits	19.2 megabits	Medium-fidelity audio
40,000 Hz	16 bits	38.4 megabits	High-fidelity audio (for humans)
100,000 Hz	16 bits	96.0 megabits	Ultrasound

How Much Information Is Needed for Audio?

How much data is required to communicate audio compared to images? This depends on the level of precision required to reproduce the original signal.

For a linear (time domain) signal, the amount of data required to transmit one second of audio is the sample rate times the number of bits per sample. CD-quality audio, which captures the full spectrum of human hearing, operates at 44,000 samples per second, with 16 bits per sample, which works out to 528,000 bits per second for a single audio channel, or about 31.6 megabits (or approximately 4 megabytes) for 1 minute. A one-minute high-fidelity (for humans) audio clip is therefore similar in size to a moderately high-resolution color image.

Compared to video content, audio requires much less information to represent, roughly 1/1000th as much, even for fairly high-quality audio streams.

Recognizing Audio

Audio samples are represented as a linear series of numbers. The number of possible sample values is determined by the number of bits used to represent each sample. A 16-bit audio sample would have 2^{16} or 65,536 possible values.

The sender can assist the receiver in recognizing audio samples by including reference signals that have a simple, repetitive pattern. The sine function, and the signal it produces, is an excellent source for a calibration signal. This function is used extensively in signal processing and is something an astronomically literate civilization would need to understand to build wireless communication systems.

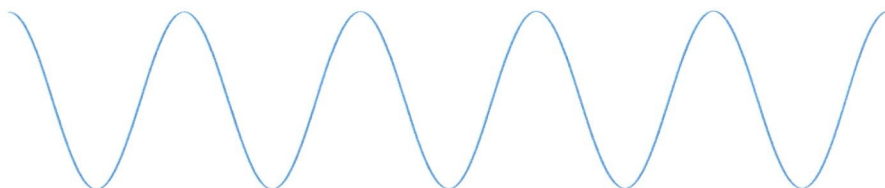

Fig. 13.2 A pure tone (sine wave) as it appears in a time domain graph. (Image credit: Brian McConnell)

Fig. 13.3 A 16-bit binary representation of the signal plotted above. Notice the repeating pattern in the bit field (left to right). (Image credit: Brian McConnell)

Fig. 13.4 A multi-tone signal consisting of four distinct frequencies ($f = 1, 3, 5$ and 7) as it would appear in the time domain. (Image credit: Brian McConnell)

Fig. 13.5 A 16-bit binary representation of the signal plotted above. Notice the repeating pattern in the bit field (left to right). (Image credit: Brian McConnell)

The repetitive nature of these reference signals should assist the receiver in identifying these as potential audio samples and from there to decode arbitrarily complex audio samples.

Defining the Time Base

The tricky part of including sound in an interstellar message is how to define a key unit of measure: time. A typical human measure is the frequency or tone of a sound in cycles per second, or Hertz, which is rooted in time. The problem then becomes how do you define a second? Our definition of a second is arbitrary. An extraterrestrial civilization will certainly measure time in different units than we do because there is no magic unit of time.

The Hydrogen Spin-Flip Transition

Fortunately, there are natural processes that operate at very precise time intervals. The hydrogen spin-flip transition, which should be familiar to anyone who is proficient at astronomy, is an example of a naturally occurring atomic clock (see the chapter Communicating Fundamental Units and Scientific Information). This clock was used by Carl Sagan et al. to define a fundamental unit of time for use in the plaques that were attached to the Pioneer and Voyager space probes. Using this process, we can precisely define a unit of time of 0.704024183763126 nanoseconds (billionths of a second).

Using this as a building block, we can define larger or smaller units of time in relation to this unit. Let's say that we want to indicate that a block of audio is recorded at 40,000 samples per second. We can express this in terms of spin-flip transitions as 35,510 spin-flip transitions per sampling interval.

Alternatively, the sender could describe the duration of an audio clip in terms of spin-flip translations. It is then straightforward to divide this by the number of samples to work back into a precisely calibrated sampling/playback rate.

Of course, there is always a risk that the receiver will not understand a metric like this, so it is a good idea to have multiple ways to work out common units of time.

Pulsars – A "Standard Bell"?

Earlier we showed how we can use mutually observable astronomical objects to calibrate image encodings. What exactly do we mean? Pulsars – rotating neutron stars that emit strong radio beams – are one candidate. They are visible over intergalactic distances and emit radio pulses at very precise intervals. The same equipment that is used for radio-based SETI searches is used for radio astronomy, so it is reasonable to assume that someone who is capable of doing SETI surveys may be familiar with pulsars.

One way the sender could call attention to this type of source is by sending several images of the star field adjacent to a pulsar and pair these with an audio sample mapped to the pulsar's signal intensity.

Fig. 13.6 An X-ray band image of the pulsar PSR J0437–4715, taken by the Chandra X-Ray Observatory. An image like this might be paired with images taken at infrared, visible, and ultraviolet wavelengths to assist the receiver in pinpointing this patch of sky. (Image credit: Chandra X-Ray Observatory)

These images could then be paired with audio samples of the pulsar's signal. The pulsar PSR J0437–4715 rotates at 173.7 times per second. The receiver could then match this audio sample against their own observations of the same object to precisely determine the audio sample rate used. This approach could be repeated across a representative sample of pulsar sources to further assist the receiver in calibrating the time base used for audio.

Environmental Sounds

Another way to assist the receiver in calibrating audio decoders is to use audio examples that are likely to be common on terrestrial planets. These could include sounds like lightning, surf, and so on. The specific tonal qualities of these will be affected by factors like atmospheric pressure, temperature, and gas mixture, but they should have a similar overall structure that makes them recognizable, although the underlying time base may be a bit off. This will at least enable the receiver to guess reasonable estimates for the time and frequency units used, even if they are unable to recognize the more precise approaches based on a shared understanding of pulsars or hydrogen state transitions.

Echolocation

Earlier in the book, we discussed the possibility that another species might see through non-visual means such as echolocation. How might they communicate "images" in a format that maps well to how they see with sound? This is a tricky problem. The internal mental image that is derived from the underlying audio is an example of *qualia*, an internal experience that may be difficult or impossible for a third party to understand.

One way to describe a system like this is to describe the audio emitted and reflected back to the viewer, as seen from several different viewing or listening points. Several species of animals see via echolocation, including bats and dolphins. When using echolocation, they produce trains of pulse that are reflected back from the objects and surfaces in front of them. A simple configuration for a system like this has an emitter that generates the sounds and multiple receivers that are spread out spatially. By working out the time required for the echoes to reach each receiver, one can calculate the shape of the objects and environs reflecting these sounds.

One way to depict this process numerically is to combine a spatial diagram of the system with audio samples for the emitter and the receivers. The reflected sounds received by the microphones will arrive at different times depending on their location and distance from the objects around them. So not only can the sender demonstrate the principle of echolocation, but they can also send working examples. One might start with a simple system and then work into more complex geometries from there.

Audio Compression

Like images, audio recordings require a large amount of information, although much less than motion picture sequences. There are a number of techniques that can be used to compress these, including both lossless and lossy algorithms. This is another example of where including computer programs in the message will be useful, as the sender can include an algorithm that implements an arbitrarily complex decompression algorithm for audio content.

A common strategy used in a number of compression algorithms is to delete faint background sounds. This typically is done by converting a time domain (linear) signal into a frequency domain signal (sonograph), which plots signal strength versus frequency over time. This can then be compressed by removing the frequencies whose signal strengths are below a certain threshold. The decompression algorithm works by converting the reduced sonograph back into a time domain signal and can be a relatively simple program (the compression step is typically more complex). This is how popular formats like MP3 work, and it can reduce the information footprint by a factor of 10- to 30-fold. This is a significant gain, so it would not be surprising to discover that the sender employs lossy compression for some audio types.

Lossy compression does come at a cost – the loss of information in the reproduced signal. In this case, low-level background sound components are filtered out, so the reproduced signal is similar to but not exactly the same as the original. If the sender does not want any information to be lost, they will not want to use lossy compression, but this may be acceptable for many types of audio samples where an approximate reconstruction of the original signal is good enough.

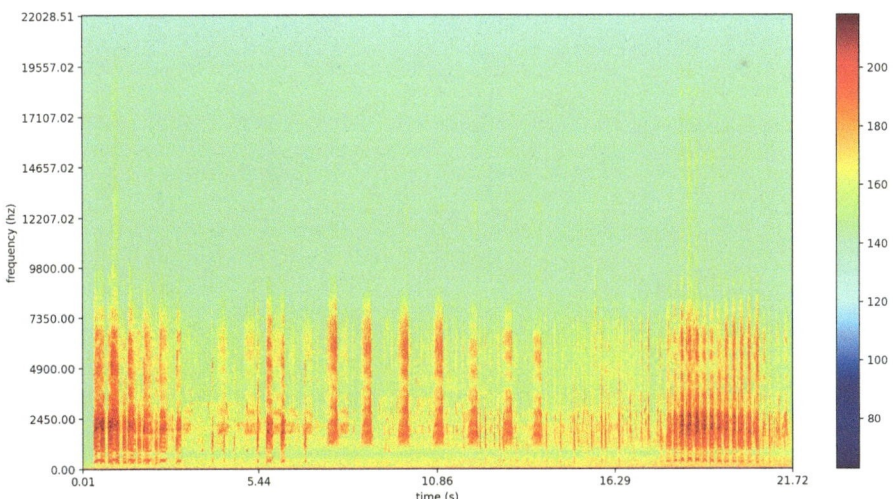

Fig. 13.7 A spectrograph of duck calls. The vertical axis represents frequency, the horizontal axis represents time, and the color represents intensity. Notice that this format enables the viewer to see how the sound is structured and that there are large areas of low intensity (the blue and green areas) – information that can be deleted while preserving the most important components of the sound sample. (Image credit: Brian McConnell)

14

Communicating Fundamental Units and Scientific Information

We've already demonstrated several examples of how mutually observable objects can be used to explain how information is encoded, as in the example of images. We can use this approach to leverage a shared understanding of natural processes and define basic units of measure for time, distance, mass, and other physical phenomena and to do so with a high level of precision.

Time

Our units for time (seconds, hours, days, etc.) are arbitrary and are derived from the length of an Earth day or year. There is nothing special about the length of a day on Earth, nor is it likely that other worlds will measure time in the same units we do.

So, how can a sender define a unit of time?

Fortunately, there are natural phenomena that occur throughout the universe that have very precise timing. The spin-flip transition in hydrogen, as we have encountered previously, is an example of a phenomenon that should be familiar to someone who is familiar with radio astronomy or quantum mechanics. This transition occurs between very closely spaced energy levels in hydrogen atoms and results in a radio emission whose frequency is exactly 1420405751.766699999967067 Hz (cycles per second). Inverting this number results in a unit of time equivalent to 0.704024183763126 nanoseconds (billionths of a second).

Once a fundamental unit of time has been defined, it is then possible to describe any amount of time in terms of that unit, just as we can define a

© The Author(s), under exclusive license to Springer Nature Switzerland AG 2021
B. S. McConnell, *The Alien Communication Handbook*, Astronomers' Universe,
https://doi.org/10.1007/978-3-030-74845-6_14

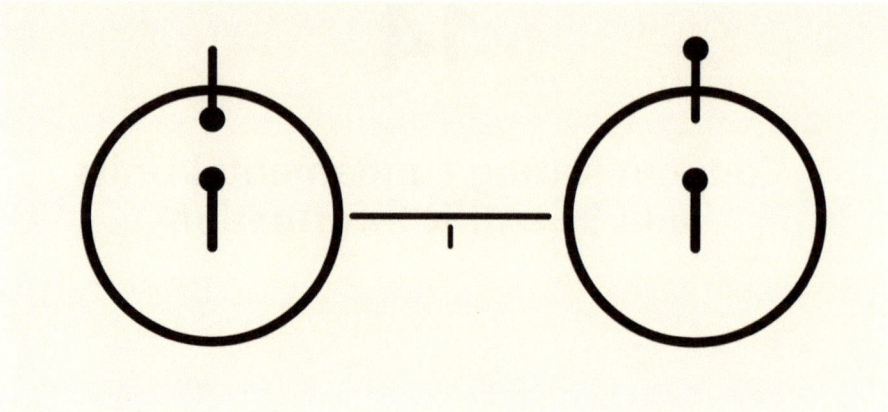

Fig. 14.1 The spin-flip transition, as depicted on the Pioneer plaque developed by Carl Sagan et al. It was used to define one unit of time as 0.704024183763126 nanoseconds. (Image credit: NASA)

minute as 3600 seconds. We can translate everyday measures of Earth time into spin-flip transitions as follows:

One second = 1,420,405,751.77 transitions
One minute = 85,224,345,106.00 transitions
One hour = 5,113,460,706,360.08 transitions
One day = 122,723,056,952,642.00 transitions
One year = 44,823,639,312,108,300.00 transitions

There are other naturally occurring clocks that can be used to define units of time, but the hydrogen spin-flip transition is thought to be an ideal choice, as hydrogen is so abundant and should be well understood by astronomically literate civilizations.

The Pioneer Pulsar Map

The hydrogen spin-flip transition was used to define a basic unit of time for the plaques attached to the Pioneer and Voyager space probes. This was then used to build a map that depicted Earth relative to nearby pulsars at the time the probes were launched.

There is a very interesting trick to how this map works. Pulsars are characterized by the rate they spin at and the radio signals they emit pulse at the same rate. Like a spinning top that loses momentum, they spin down over time at a predictable rate over long periods of time. The pulsar map on the plaques doesn't describe Earth's distance to each pulsar, but rather the pulsar's spin rate as seen from Earth at the time of launch.

Imagine then that several millions of years from now, someone picks up one of the Pioneer probes and compares the spin rates described on the plaque against the same pulsars as they see them at that future time. They will be able to work backward in time to figure out where the pulsars were at that time and, from that, calculate pretty precisely when and where the probes were launched.

The pulsar map has since been updated by Dr. Nadia Drake and Dr. Scott Ransom, who have developed a new version that references millisecond pulsars. This class of pulsar, which had not yet been observed when the first map was created, is much longer lived and should still be observable in hundreds of millions of years. The pulsars referenced in the Golden Record spin much more slowly and are expected to fade out in a few million to tens of millions of years, so the Golden Record will outlast the pulsars used in the original map.

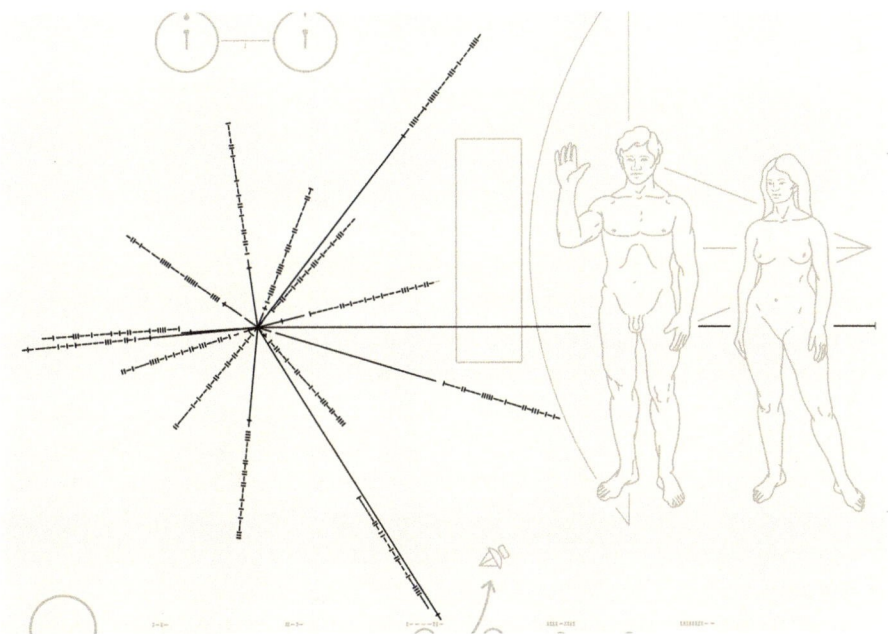

Fig. 14.2 A diagram of the Pioneer plaque with the pulsar map highlighted. Each radial is annotated with binary notation that depicts the radial's pulsar's spin rate in terms of hydrogen spin-flip transitions. (Image credit: Wikimedia Commons)

Distance

Next we'll want to define a basic unit of distance, which can be done through direct examples, by pointing to objects with well-known dimensions and by defining distance in terms of velocity divided by time.

The speed of light, c, is a known quantity and is the same everywhere. One method of defining a basic unit of distance is to express this in terms of c times the basic unit of time. Using the unit of time derived from the hydrogen spin-flip transition, we get a basic unit of distance of 0.2110611405418 meters (or roughly 21 centimeters), and since this is known to a high level of precision, this is a useful reference scale for things that range in size from atoms to cosmological scales.

If one wants to define a basic unit of distance at a much smaller scale, they'll probably want to refer to a common subatomic particle such as the proton and use that to reinforce other metrics used at larger scales. It's important to note that referencing more than one way to measure a unit of distance doesn't undermine the others. Instead, it provides the receiver with multiple paths to understanding the concept of distance.

Mass

To describe a basic unit of mass, there are plenty of mutually observable objects we can point to whose mass will be the same everywhere. Again, hydrogen is an excellent choice to work with because it is ubiquitous and will be well understood by anyone who is proficient at astronomy and physics. The rest mass of a neutral hydrogen atom should be a good unit of mass, especially for defining the mass of larger objects.

Another option is to define mass in terms of energy, which we could do by translating the energy of a 21 cm photon emitted by a spin-flip transition, which we previously used to define a fundamental unit of time, into mass. This works out to about 1.05×10^{-41} kilograms. The approach of using mass-energy equivalence would also indicate that the sender understands this aspect of relativity.

As with distance, the sender can define basic units of mass in terms of many objects, as a collection of references like this will provide multiple paths to comprehension, as well as greater accuracy.

Energy

A basic unit for energy can likewise be described in terms of mutually observable processes. Well-known transitions, such as the spin-flip transition, are good candidates since they release precise amounts of energy in the form of photons with precisely defined wavelengths. The energy released in the hydrogen spin-flip translation works out to about 9.4×10^{-25} Joules. Processes like this are attractive because they always emit or absorb a photon, whose energy can be calculated as a function of its wavelength.

Electric Charge

A unit for electric charge is also something that can be defined in reference to mutually observable objects. This could be done in a number of ways, but since we used hydrogen in other examples in this chapter, let's look at how it can be used to refer to a basic unit of charge.

One way to do this is to depict a hydrogen atom with and without an electron shell. The net charge of a neutral hydrogen atom will be 0, while the charge of an ionized hydrogen atom (a naked proton) will be +1. This example could be extended to depict several different atoms with and without their electron shells to reinforce what the sender is getting at. Larger units of measure that are useful at practical scales can be defined in terms of these atomic scale units.

> ### Thoughts on Basic Units and Notation
>
> Whether we are creating a primer for an outbound message or parsing the primer of an alien message, the basic units of measure that would be defined are not intended to be "user-friendly." We are used to dealing with metric or imperial units of measure that, when they were first defined, had no connection to the basic physics underlying the universe. Like our measurements for time, they were subdivisions of familiar units of measure that we had agreed on by consensus over millennia.
>
> What an author will be doing in a scientific primer like this is to define basic units of measure in relation to fundamental objects or physical processes that take place at the atomic level, and that can be measured very precisely. The resulting fundamental units are generally very small and are therefore a bit unwieldy in describing larger objects or processes. That's okay: if needed, it is always possible to build equivalence tables that define larger units of measure in terms of the fundamental units.

15

Semantic Networks and Constructed Languages

© The Author(s), under exclusive license to Springer Nature Switzerland AG 2021
B. S. McConnell, *The Alien Communication Handbook*, Astronomers' Universe,
https://doi.org/10.1007/978-3-030-74845-6_15

For most of human history, language was a spoken phenomenon. Only recently in cosmic terms have we developed systems for written communication, which has enabled us to communicate over long distances and to preserve information over long periods of time.

Our written languages are mostly derived from spoken language and shared experiences, so it is very unlikely that a natural language or its written form will be mutually understandable with an alien species. We should not expect anything like spoken language in interstellar communication, except maybe when providing examples of animal communication.

On the other hand, one can communicate a great deal using sets and symbolic networks and can do so using a minimal amount of data. This sort of symbolic communication system also forms the basis for more abstract language, with common mathematical terms and logic as its foundation. This is known as a *constructed language* or *interlingua*.

Each image, sound, or other representational object can be thought of much like a file on a computer. As the number of items in the collection grows, it's helpful to have a system for organizing items, making them easier to find, and linking them to each other via logical associations. This is typically done with metadata – data about the data it is describing – which we discussed previously in Chap. 7 on Lessons from Computing and Communications. While the methods vary, a good implementation will allow the author to describe and classify the object to make it searchable and to describe how it relates to other objects and collections in the system.

This system can be extended to connect representations of objects to abstract ideas. Just as something similar to a file system can help organize representations of objects (e.g., images), it can also be used to catalog ideas. This is known as a *semantic network*.

The basic idea in a digital semantic network is that each unique concept or idea has a unique numeric address. The number assigned to the concept itself has no meaning; its purpose is to give the concept a unique identifier, much like each device on a network has a unique network address similar to a phone number.

Combining an address space for ideas with a set of symbols that describe the relationship between any two ideas in a network enables the sender to build up a complex graph of relationships using a simple set of basic connections, between not just abstract ideas but also representations of those ideas (pictures, sounds, etc.). Ideally, what you would like to have is a system that enables you to describe the following for each idea or object representation in the system:

A symbol for equality (A is the same as B)
A symbol for inequality (A is not the same as B)
A symbol for similarity (A is similar to B)
A symbol for dissimilarity (A is not similar to B)
A symbol for contains (A contains B)
A symbol for set membership (A is a member of B)
Symbols for relative degree (A is greater than B, A is less than B)

The basic idea is to describe concepts such as set membership, similarity, dissimilarity, degree (size), and so on. These relationships map to a small vocabulary of symbols that can then be applied to a graph that connects different concepts, as if they are mapped in a network as shown in Fig. 15.1.

Unicode for Ideas

Unicode is a character encoding system that was developed in the early 1990s to provide better support for non-Latin alphabets on computers. Up until that time, most computers and software were built around the ASCII character encoding system, which was really only designed to work for American English. Support for other alphabets, and especially non-Latin alphabets, was inconsistent at best and nonexistent for many writing systems. Unicode solved this by creating an extensible encoding that allowed for over one million unique characters, enough to accommodate every letter or symbol used in every alphabet. Unicode, and its cousin UTF-8 encoding, have since gone on to become the dominant encoding system used on the Internet.

What does this have to do with semantic networks? A semantic network will use a similar code space to map unique ideas or words to unique addresses. The problem is similar to that faced by the designers of Unicode, but even greater in scale. How many unique ideas or words might exist among a long-lived network of civilizations? Millions? Billions? More? How do you create a system that allows newcomers to coin new meanings and do so without stepping on other users of the system?

The trick is to create a large address space, big enough to provide unique addresses for every unique item in the system and also big enough that two users who randomly claim an address to use for a new idea are unlikely to use the same one (an *address conflict*). Fortunately, the number of addresses scales exponentially with the number of bits used in each address, by 2^n. A 32-bit address (just 4 bytes) yields over 4 billion unique addresses. A 48-bit address yields over 281 trillion unique addresses. A 64-bit address yields 18 quadrillion unique addresses. The number of possible addresses expands astronomically

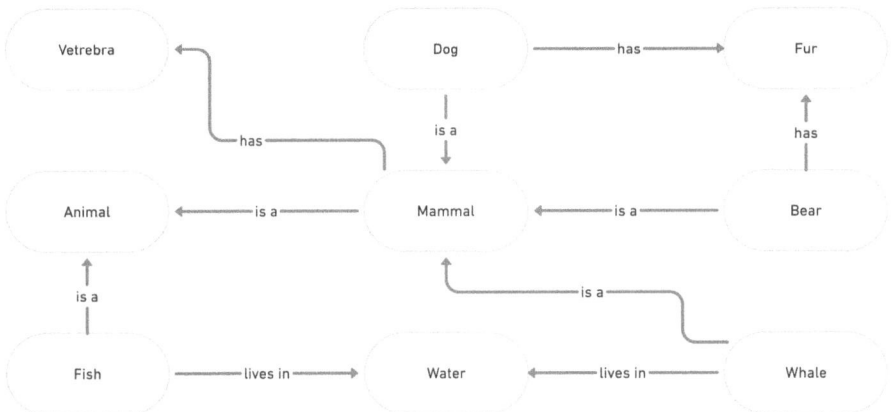

Fig. 15.1 An English representation of a small semantic network that describes the relationships between several types of animals. (Image credit: Brian McConnell)

from there. At 256 bits, the number of addresses is comparable to the number of atoms in the observable universe.

The trick in designing a system like this is mostly about avoiding address conflicts. Unlike here on Earth, there won't be a Unicode consortium to mediate a dispute, and even if there is, the Galactic Internet Council might take several hundred years to resolve it. An easy solution is simply to define an address space that is so large that the odds of two parties randomly choosing the same address are vanishingly small.

Let's consider how one might communicate the idea of planets and moons in an interstellar message. The sender could do this by including a set of photographs, some of which contain photos of planets, some of which contain photos of moons, and some of which contain photos of both. Each photo is labeled with a semantic network expression that describes set membership (e.g., "This photo contains 'moon'"; "This photo contains 'moon, planet'"; and so on). The recipient can then use a solve for x pattern to work out which symbols correspond to which concept, as well as the relationships between concepts. See Fig. 15.2.

How will metadata like this be represented? There are several general ways to do this and a limitless variety of specific implementations. One approach is to embed the metadata within the object representation itself.

An advantage of embedding metadata within each object representation is that information about what the item represents and how it is connected to other objects in a collection always travels with the object representation. This is not unlike adding labels to cue cards. The main cost of this approach is that extra space is required in the image, audio clip, etc., to contain metadata, and this may initially be confused with the primary object payload.

Fig. 15.2 An example of an object representation (image) with metadata embedded within it. Notice the binary code in the upper left corner. This is an example of how metadata can be embedded in an image as a sort of cue card

By tagging object representations like this, the sender will be inviting the recipient to map the differences in codes associated with individual images. The example image above has four codes for "planet," "water," "ocean," and "water ice." Images of other planets and moons would be tagged with different sets of labels defining their characteristics. By studying the differences and similarities between object representations and the coded labels associated with them, the recipient will often be able to deduce the meaning of these codes using a solve for x pattern.

The point of these examples isn't to predict that an ET's semantic network will be constructed in a specific way, just that these are patterns to be on the watch for. They are straightforward to implement and will not get in the way of the receiver learning how to comprehend basic object representations, even if it takes time to understand additional layers of information.

This type of system, although it is artificial, could also satisfy the seven requirements for a language described in the chapter Animal Communication Research. One can think of it as a machine-readable interlingua that can be converted into the user's native language, at least to the extent that concepts defined in the semantic network can somehow be mapped to the user's language.

Here it is important to revisit the idea of communicating observables versus communicating qualia. Objects or actions that are tied to physical processes should be straightforward to understand, while concepts related to internal states of mind or experiences might prove difficult or impossible to grasp.

Learning to Parse the Network

The first step in decoding a semantic network and expressions that utilize it will be to learn the statements that are used to link addresses within the network. In their simplest form, they will take a form like:

```
Node A        Operator      Node B
```

The order in which the elements are listed may vary, so other variations on this form will be:

```
Node A        Node B        Operator
Node B        Node A        Operator
Node B        Operator      Node A
Operator      Node A        Node B
Operator      Node B        Node A
```

While the number of nodes in the network may be very large, the number of operators used to link nodes should be considerably smaller. Some operators, such as those that define set membership or similarity, should be very common.

The network is then defined via a large set of expressions that describe how nodes are related to each other. We might see a series of numeric expressions like this:

```
180052    229004     7
347414    382206     171
292490    1003474    144
488939    908629     201
931782    488939     144
188862    246339     136
735893    17620      224
878788    1351       220
```

990062	103475	215
863976	998094	193
101050	819053	7
665465	582165	144
71860	981701	195
362224	813023	214
208482	644831	224
904643	77088	26
965857	933017	171
130639	905266	214
582165	867645	26
1351	909832	26
517721	349920	195
871975	672245	26
905266	447085	136
428709	325563	144
315385	532049	215
242532	599743	7
677119	965857	214
857601	995735	220
917413	696078	193
528626	509077	220
17620	208482	201
625545	904643	26
720387	814012	7
68763	101072	201
247629	17620	195
979082	120297	144
663814	725755	144
599743	252015	142
857345	619506	224
532049	325563	26
908629	936525	195
216775	478881	193
558384	271692	161
349920	160853	26
776577	137970	193
540724	769946	224
129469	103475	214
912789	521701	215
644569	63363	224
247413	292490	136
931772	76473	193
796959	5184	193
951164	878788	224
573871	1042739	201

```
611276      959828     220
813023      582800     7
1039280     612964     220
```

At first, we would not know what any of these numeric "words" mean. However, we would notice some patterns. In this example, the number of values encountered in the third column is much smaller than the number of values encountered in the first and second columns. Because these numbers show up more frequently, we'll want to focus on deciphering their meanings first.

As we learn what the operator codes represent, we can start to build out a graph of the network and map how nodes within the network are related to each other. Once we have decoded the meaning of frequently used operators, we can then build computer programs to index the network and run queries against it.

Querying the Network

The expressions that define how nodes in the network are linked to each other can be read into a database that can answer queries like "Which sets does concept X belong to?"

We can then write programs that enable us to query the semantic network and explore how different concepts are related to each other. Even before we learn the meanings attached to nodes, we can begin to understand how they are grouped and connected and can also identify regions of the network that are islanded or isolated from other parts of the network.

What Nodes Is Node X Connected to?

In this query, we want to know which nodes a specific node is directly or indirectly connected to. This query will tell us which nodes a node is connected to and how they are related. This can be presented to the user as a program or function that takes the form:

```
list_of _links = who_am_I_connected_to(node_address)
```

This function would return a list of nodes the queried node is connected to and how. The function will return a list of statements like:

2712 is similar to 6712 which is contained by 9281
2712 is contained by 10932
2712 is unequal to/opposite of 45192

Even if we don't know the meaning of individual symbol codes, we can learn which codes are most connected to others. More links equals more paths to comprehension and therefore a place to focus our efforts).

How Is Node X Related to Node Y?

Here we want to understand how two nodes are related to each other (if a link path exists). This query can be presented to the user in the form of a function probe_relationship(node_a, node_b). The function will return a list of expressions that connect the two nodes or NULL/NONE if no path exists to link the two nodes.

The number of links and the directness of the path between the two nodes will hint at how related or connected they are. For example, if Node A is directly connected to Node B with an equivalency operator, we'll know that if we can learn the meaning of either, we will learn the meaning of the other.

For other nodes, we might find that there is no direct connection or only a very indirect connection between them. The sender might describe a set of experiences that are closely related, but have no direct connection to observable processes that we can understand. When displayed as a graph, these concepts would appear as a disconnected group of nodes or island. In order to understand them, you would need to understand at least one of the concepts on that island, and from there work out the meaning of the others.

Which Nodes Are Most Connected?

Nodes that are highly connected to others, especially in terms of set membership, may offer many paths to comprehension and will be a good place to focus our efforts prior to taking on nodes that are only weakly connected to others.

The sender might choose to include a large corpus of planetary images of the planets and moons they have surveyed or are aware of. The images would be linked to nodes in the semantic network as follows:

Each planet or moon has a unique identifier in the semantic network. Think of this as a numeric name or record locator

Every planet and moon is linked to the node for the concept of a spheroid

Each planet's node is linked to the node or identifier for the concept of a planet

Each moon's node is linked to the node or identifier for the concept of a moon

Individual images are labeled with links to the nodes describing attributes of the scene depicted (includes water, includes water ice, includes continents, etc.)

When we first examine this segment of the network, we won't know what any of the numeric codes represent, but by querying the network, we will learn how they are connected and will notice some patterns:

Every image of a planet is linked to a specific, unique code

Every image of a moon is linked to a single code

Each image is linked to one or several of a small list of codes

By comparing images and what nodes they are linked to, we can use a solve for x pattern to associate numeric codes with human language terms. For example, we might notice that images containing worlds with ice caps are always linked to node 1927, which we learn maps to water ice. By spotting the similarities and differences among these samples and what they are linked to, we can build a table that maps numeric codes to words we understand.

Linking a Semantic Network to Other Media Types

A semantic network that references only itself will not be particularly useful, but when it is linked to other representations of objects or processes, it will be possible to develop a rich vocabulary of concepts and their relationships.

Any media type that can be conveyed by the communication system can be linked to a node in a semantic network. This is a potential pathway to comprehending more complex or abstract concepts that can't be expressed in a purely mathematical form. Let's look at a couple of examples.

Turbulence

As was briefly mentioned in Chap. 9, turbulent flow is an example of something that is easy to understand via observation, but is not at all easy to describe in purely mathematical form.

A sender wishing to define a symbol for turbulence may send a collection of images, such as Fig. 15.3 below, or simulations that depict turbulent flow and link them to a semantic network node that represents the concept of turbulence. Other concepts can, in turn, be described in terms of this root concept for turbulence.

Fig. 15.3 An image of a plume from a candle that transitions from smooth to turbulent flow at the top of the image. (Image credit: Wikimedia Commons)

Nonlinear Sensitivity (The Butterfly Effect)

Nonlinear systems often are extremely sensitive to starting conditions, such that a small difference in starting conditions can lead to very different outcomes as the system involves. This is often referred to as the *Butterfly Effect*, in reference to the action of a butterfly flapping its wings, which later leads to large-scale differences in weather far in the future.

Let's imagine that the sender wants to describe a symbol for this sensitivity. They could do this using sets of positive and negative examples.

They could describe positive examples using algorithms that model nonlinear systems and, by varying starting conditions by small amounts, point out that they are sensitive to these differences. The output from these algorithms would be linked to the node for this concept using a "belongs to set" operator.

They could describe negative examples using algorithms that model linear systems that are not sensitive to small changes in input conditions. An example of this would be a function that calculates the sine of an angle. The output from these algorithms would be linked to the node for nonlinear sensitivity, except using a "does not belong to set" operator.

The Limits of Communication

A semantic network will be capable of mapping any concept that has a distinct meaning and that can be defined in relationship to other concepts. That doesn't mean that someone studying the network will be able to comprehend every concept described within it.

Let's imagine that we are composing a message to be transmitted to another civilization and want to include a lexicon of human emotions and how they are related. To do this, we add a few nodes to the semantic network and describe how they are connected. We'll start with the following entries:

18072	brain
19025	brain state (mood)
19412	happiness
19817	sadness
29154	anguish

And then map them to each other as follows:

19025	136 (is a property of)	18072
19412	26 (belongs to set)	19025
19817	26 (belongs to set)	19025
29154	26 (belongs to set)	19025
29154	7 (is greater than)	19817
19817	144 (is opposite of)	19412

A naive receiver would be able to recognize these as distinct concepts and would be able to see how they are related to each other in terms of degree and set membership, but they might have no basis for understanding what a mood is. Humans are able to communicate in depth because we have a shared experience or consensus reality around which we can build vocabularies. This may not be the case with aliens, who have different biology or are not biological at all.

An interesting aspect of a semantic network is that we can map not just what we have learned but also what we do not know. Unknown subject areas will be displayed as disconnected islands in the larger network. Using the example of emotions, the reader will know how the symbols map to each other, but they will be islanded from symbols they understand. While they won't know what these symbols mean, they will know that they don't yet have a path to comprehension, or in other words, they will know what they don't know.

Building an Interlingua

Besides classifying how concepts are related to each other, a semantic network can also serve as the foundation for a machine-readable interlingua or intermediate language, a sort of ET Esperanto. The resulting language will be more like a programming language than a written or spoken language, but that's actually a good thing because the author will be forced to be precise in how they assemble symbols from the semantic network into larger expressions. Information about how words are used in natural language is often implicit and can also depend on other words or phrases in a larger expression. For example, the English word "like" has many meanings, and which one is intended is dependent on how it is used in a sentence.

An interlingua based on a semantic network will benefit from having the following features:

A mechanism for grouping symbols into larger expressions (phrases, sentences and larger collections)

A mechanism for describing and modifying noun phrases that describe objects

A mechanism for describing and modifying verb phrases that describe actions

A mechanism for describing time or tense (past, present, future, etc.)

A mechanism for describing causality (what or who is acting on what or whom, certainty, etc.)

First, we need some sort of marker to indicate how symbols are organized into larger collections or data structures. What you are looking for is something that operates like open and close parentheses. This type of operator allows a linear sequence of symbols to be organized into arbitrarily complex structures and also allows the author to precisely define how an expression is organized. In English, we use punctuation marks to do this.

With these building blocks, the author can compose larger expressions that describe how objects and actions take place, how they interact, and how they are displaced in time. These are the basic requirements for language as we think of it. A system like this also allows the author to assemble discrete symbols into arbitrarily large and structured expressions that are analogous to sentences, paragraphs, etc.

Verb-like clauses could also include symbols that describe time or tense (whether something happened in the past, is happening now, or will happen in the future), as well as symbols that describe a degree of certainty (is definitely occurring, might occur, etc.). In natural language, this information is often encoded in the verb phrase, which is spelled or conjugated differently depending on tense, plurality, and certainty.

In a constructed language based on a semantic network, there need not be a notion of conjugation; instead, one can group symbols into a verb phrase that contains the information needed to convey meaning, such as:

```
((symbol:climb)(tense:future)(certainty:definite)(usage:verb))

   NOTE: this example is expressed in English form for
readability
```

A program that parses this verb phrase can transform it into a more user-friendly form, and because the structure and meanings in the expression are defined precisely, there will be little ambiguity in this process. This verb phrase can, in turn, be combined with other phrases to form a larger expression, similar to a sentence.

The examples so far use simple expressions to link nodes, such as "A belongs to set B." These definitional links could be more complex than that and function more like sentences. This would allow concepts to be defined even more precisely in relation to each other, similar to how words are often defined in dictionaries. We might struggle with more complex definitional links at first, but if they are combined with simpler expressions in terms of set membership or degree, we should have a good starting point for probing how concepts are related to each other within a network.

Translating an Interlingua into Human Languages

An interlingua that is based on a semantic network can be translated into human languages via a straightforward process that will be easier than translating between human languages. By assigning unique identifiers to each concept or usage, a semantic network will have a built-in mechanism for disambiguation. Human languages, on the other hand, contain words that have multiple meanings in different contexts. This often leads to poor translations, not just by computer translation systems but also by professional translators.

A translation program that reads these programmatic statements out into human languages might not produce appealing prose, but it should be understandable. This type of program would work by substituting numeric addresses with human-language translations where they are known. The output of the program might read more like a markup language, similar to XML (Extensible Markup Language), but that's okay, as this can be passed through additional programs to produce more finished prose if desired. At first, there will be many unknowns, so we might see output like:

```
   ?? (718226) is planet (5721) has water ice (1927) ?? (62172)
?? (723109)
```

The unknown/unmapped concepts in output like this would tell us what we don't know and invite us to use solve for x patterns in working out what the undefined concepts represent. How successful we are will depend on the information being conveyed and how well it maps to our knowledge and experiences. If they are communicating mostly about observable objects and processes, we may be surprised at how much we can understand. On the other hand, if they are communicating mostly about internal states and experiences, we might have no common basis for understanding and be completely flummoxed.

Analyzing expressions that are built on this interlingua can be done in a manner similar to word-level entropy analysis in human texts. Each unique ID can be treated like a unique word, and these can be fed into an entropy analysis program (see Chap. 8 on Entropy) that measures nth order entropy throughout the collection. A large corpus can be broken down into smaller blocks that are passed into this utility, which identifies the most frequently used "words" and n-grams (collections of 2, 3, or more words/symbols). This analysis will not magically translate anything, but it will enable analysts to figure out which concepts are being referenced the most in a given part of the collection. We can match this information up with a list of the most connected symbols and work on the symbols that are used most often and have a lot of connections to other nodes in the semantic network.

16

Genomic Information

Fig. 16.1 The 16S region of human ribosomal DNA plotted as a tile grid, with one color for each DNA base pair: DNA sequence courtesy of the European Bioinformatics Institute (Homo Sapiens 16S Ribosomal DNA Sequence, European Bioinformatics Institute, https://www.ebi.ac.uk/ena/data/view/Non-coding:GU733719.1:1671..322 8:rRNA). (Image credit: Brian McConnell)

© The Author(s), under exclusive license to Springer Nature Switzerland AG 2021
B. S. McConnell, *The Alien Communication Handbook*, Astronomers' Universe,
https://doi.org/10.1007/978-3-030-74845-6_16

We should anticipate that an ET transmission might include genomic (genetic) information from organisms on their world. On Earth, the information encoded in DNA can be transmitted as digitized information, where each base pair encodes 4 possible states or 2 bits of information. There are a number of reasons why an ET civilization might choose to do this, which will reveal information about how life developed on their world, how it functions, and how it did or did not spread to other worlds. This information would have the potential to answer some big questions, some of which we could not get the answers to short of traveling to other star systems.

Why Include Genomic Information?

At ET civilization might have a number of reasons for including genomic information in their transmissions or inscribed matter probes, as would we.

Comparing different life systems: a long-lived network of civilizations will be able to compare notes about how life functions on different worlds. By sharing genomic information, they will be able to answer big questions about how life differs from world to world and whether there are interstellar delivery mechanisms such as panspermia at work.

Evolutionary history and population dynamics: genetic information is passed along as species evolve and branch off from each other, and by studying how genes change over time, one can learn about the history of an organism.

Preservation: no matter what happens to a civilization's home system, they would be able to preserve a record of life from their world by doing so. At the very least, other civilizations would be able to study and learn from the evolution of life on their world. This would also leave the door open to their life or biosphere being reconstituted on new worlds.

What Might a Primer Look Like?

Let's start by looking at a human gene sequence in binary code. The code shown in Fig. 16.2 is from the 16S ribosomal DNA sequence and is represented in binary form. In this sequence of digits, the nucleotide adenine is represented as 00, cytosine as 01, guanine as 10, and thymine as 11.

If you did not know you were looking at genetic information, you would have no way of knowing that based solely on the information content. The numbers appear to be more or less random in distribution. The sender will need to make it clear how genetic information is represented throughout the transmission. To do this, they will need to include a primer.

Early experiments at interstellar communication, such as the Golden Records flown with the Voyager space probes, included information about DNA and how it encoded information, although they did not include genetic sequences. They did not have the capacity to transmit much information, and genome sequencing was years in the future at that time. Let's look at how the Voyager records describe genetic information, as that is a base that can be expanded on if one has more information to work with.

```
10011100000000101110010010101010000000010101000111010100010111110001110001010010000010000010111
11001001010000000010100111111000101010000000110000001011001100101001100010010000000001111100000
00010111101001100100001100100011001100101100010110010000010101000000010001110000000000000111100
11000001010000100100110000110011001001000001010000011100000010101011100110001011111011110010011
00001110000011110000011100100000000110000011111110010000101000100010010100000001001110000010000
01010101011000000001010010000110001001110001011100001000000100100111000000001000100100010001
01011011011100111011001001000000001100101110101000001000111111001100101011001000101001100001
00000001011100010110001001011110101110011001001110101011101101010100001000100010000011011111
00101111010000011111110000001111110010101000100010000001010111011100000011010101011111101100
00001111110000011110111100101101010000000100010100000010010011101111111010000100011100101000
00000000000101111110110010001000100010011000000000000111110000010001010100110010110010010101
11000000000100100100101000101000011110000100000001001101111010000100110100000100010101000111
00010111000000000011010101000000010011001100000111000000011101011101000100010101000011111010
00010100001101110011001001100000010011001101111001110010110111110010110110010010000010110000100
11100000000001001110101101110101100100110000100101110011001101001000111100000000000011110
0000111100001000011100000010010010101000011001101110001000011010000010100000100001011010011
11001110001010111010001111011010000010101000000010010010100100111001110100110001010000000010
101111000000000000001011000000001010000001101101010001000000011011111000101010110010111110111111
000101000000000001001101000010111011100100100110100010100101100111100100010100100010110010111
10010010010011000010001001101111111000000101001011001001011000101011000010110111001001001001111
10101001001001100001010100011111011110110111110001111111110000001010010110000000001111100011
11010100010000101010111101000100111101101101101111100011111111100000101000101110000000011110001
0111100101011011100000100010100101010100100110000100010010010100000100001100010000010000010101
00111010001001111110000111111001111000011100100000001001011000101110000010000000010101000100
1010110101110000000111000101000000010111100100111000000000011111101101011111010101001100001
01110110100100100100000010101000001011100010010010011001010011001100111000001111000001111100001
010010110100000001001000011100011001100010011010000111100010100010000010000011111000010100
000110100000010000101110001010101010010101001100000100100110010000110101100111101110010001010
110101001100110100000010000110010101010111110001100001011011000110111110100110100100100100
110101011000110101110010010010110011001111100000010101110110110111111101111010000001100011110
0000101101011100011011110001101111000101110100100001011010001010000110101001010110110101111
1101110011011100011110100000011110101110101011101110100001000000101000010001000100000001100
001010010011100011110100010000000010011001011111010101010011011000000111000110011010001101110100
0001111100101100111100110000101010000100010101000010101000010000001001010101111
```

Fig. 16.2   A sequence from the 16S region of human ribosomal DNA. Each base pair is represented as a two-digit binary number

The diagram in Fig. 16.3 was included as one of the images encoded on the Golden Records. This is a line drawing that depicts DNA's double helix on the right and the nucleotides adenine and guanine on the left. The nucleotides are depicted as two-dimensional diagrams that show the bonds between the atoms that make up each molecule. The DNA molecule is marked up with symbols that point to each nucleotide, and ongoing annotation hints that it is being used to encode information. This diagram packs a lot of information into a small information footprint, which was a necessity given the limitations on how much could be sent via these records.

Let's reexamine this in the context of a transmission that can deliver much larger amounts of information, so the author is less constrained in terms of how much information they can use in building a primer.

We'll start with the nucleotides that comprise the basic alphabet used in DNA: adenine, cytosine, guanine, and thymine. For now, let's assume that there is an address space for symbols that we can map these to, which will allow a more complex representation of these chemicals to be mapped to

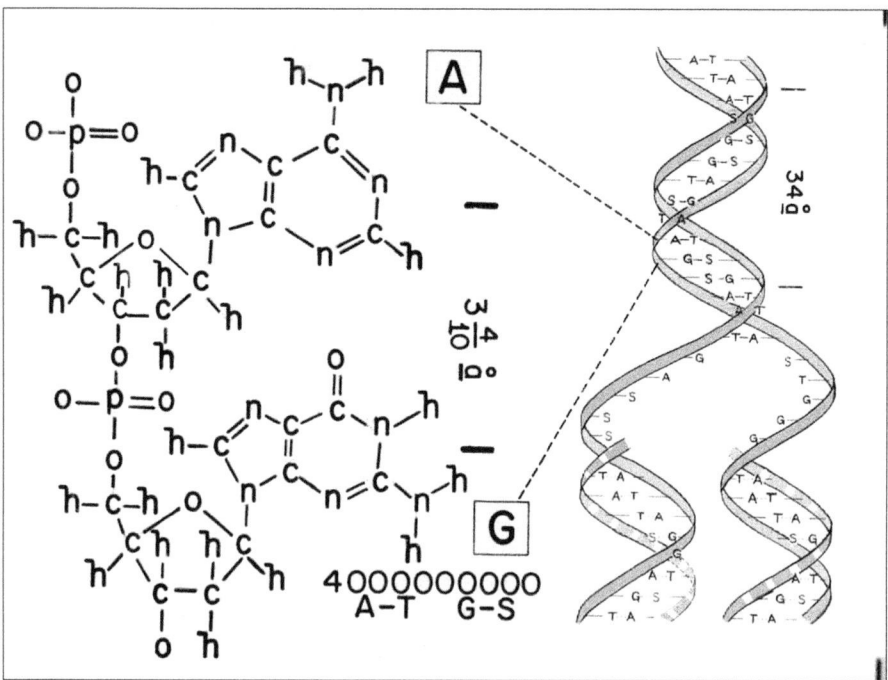

Fig. 16.3 An image from the Golden Records that were attached to the Voyager space probes. This image depicts DNA and the nucleotides adenine and guanine. (Image credit: Jon Lomberg)

shorthand notation or numeric addresses. But first, we need to teach the receiver to make the connection between these different forms.

Using the principle of providing multiple paths to comprehension, we might map several representations of a guanine molecule to a shorthand or numeric symbol that can be used in genetic sequences. The Voyager records used two-dimensional diagrams based on the notation widely used by chemists. These diagrams depict the individual atoms in a molecule and the types of bonds between them.

The diagram in Fig. 16.4, while convenient for humans to work with, has a number of features that could confuse someone else, including the implicit location of carbon atoms, which are omitted from the figure, as well as the alphabetic notation for the atoms themselves. We can make this more explicit with something like Fig. 16.5 or Fig. 16.6.

The stylized diagram above improves on conventional molecular diagrams by eliminating alphanumeric notation and by replacing it with binary references that are rooted in basic physical attributes such as atomic numbers. One important feature is missing from this diagram: the three-dimensional shape of the molecule, which will better explain how it fits with and binds to other molecules, such as its complementary nucleotide.

This can be addressed by including three-dimensional models of molecules, which, as we discussed in Three-Dimensional Images and Models, can be done in a number of ways. These could include stereoscopic images of a 3D model taken from different viewpoints, as well as 3D models defined using point clouds or similar methods. If the transmission includes algorithms, it could include a simulation that models these molecules in even greater detail. The point is that the author of the primer can include several different representations of the same molecule and then equate these with a shorthand symbol or numeric code that is used elsewhere to refer to that molecule.

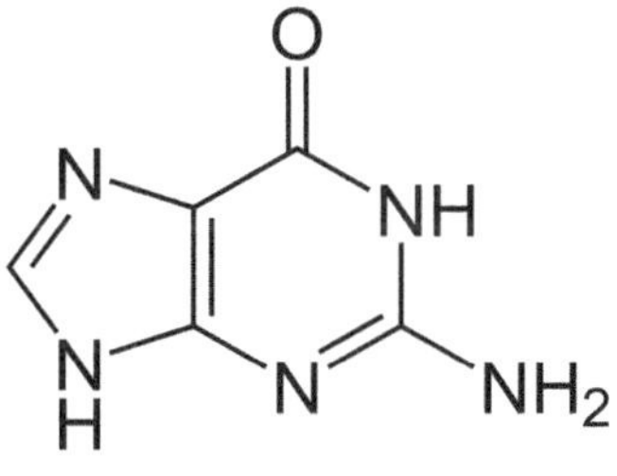

Fig. 16.4 A two-dimensional diagram of a guanine molecule (note that the positions of carbon atoms within the diagram are implicit). (Image credit: Wikimedia Commons)

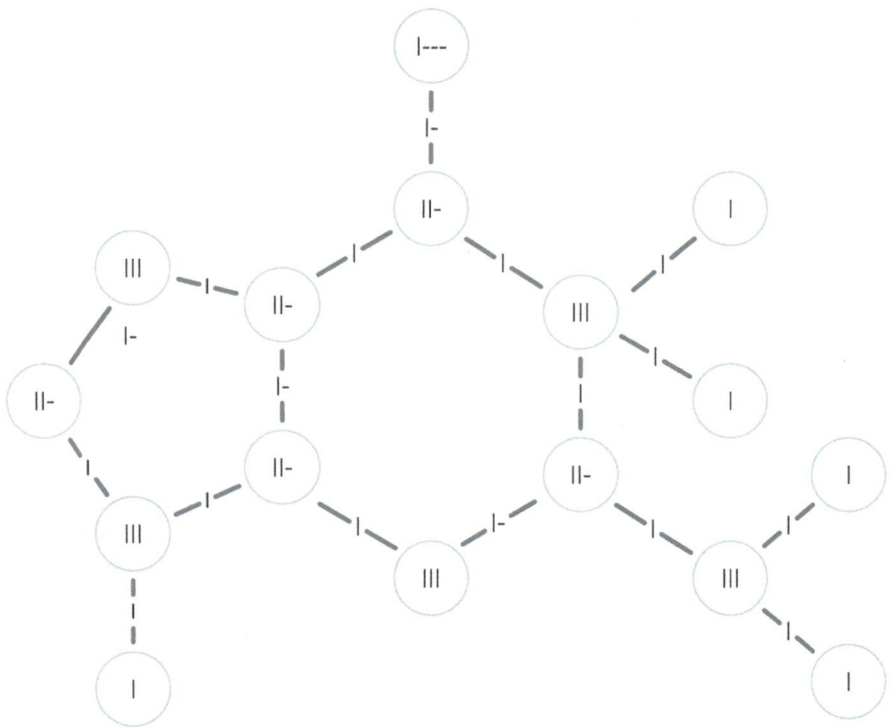

Fig. 16.5 An example of a molecular diagram that references each atom by its atomic number in binary notation (hydrogen = I (1), carbon = II- (6), nitrogen = III (7) and oxygen = I--- (8). The position of the carbon atoms is made explicit. The number of electrons shared in each atomic bond is also given in binary notation. (Image credit: Brian McConnell)

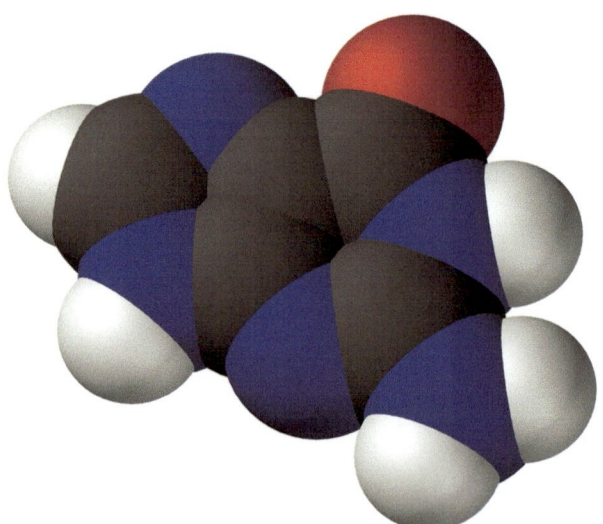

Fig. 16.6 A 3D rendering of a guanine molecule. Black represents carbon, blue represents nitrogen, red represents oxygen, and white represents hydrogen. (Image credit: Wikimedia Commons)

We would then need to repeat this process to define the other nucleotides, as well as the sugars that make up the backbone of the DNA molecule. This will enable the receiver to understand what the component molecules are and how they fit together. However, still more information is needed to explain how a chain of these molecules code for amino acids that can in turn be chained together to form proteins.

In Earth DNA, these bases are read as triplets, which results in 4^3 or 64 possible combinations, enough to code for the 20 amino acids found in all organisms, as well as codes for START and STOP, which indicate where a block of code used to assemble a protein starts and ends, as shown in Table 16.1. This triplet structure is interesting because most it also serves as a form of error correction, not unlike the n-modular redundancy code we discussed earlier.

Table 16.1 A list of base pair triplets and the amino acids for which they encode

Amino acid/operator	Codon
Alanine	GCT, GCC, GCA, GCG
Arginine	CGT, CGC, CGA, CGG, AGA, AGG
Asparagine	AAT, AAC
Aspartic acid	GAT, GAC
Cysteine	TGT, TGC
Glutamine	CAA, CAG
Glutamic acid	GAA, GAG
Glycine	GGT, GGC, GGA, GGG
Histidine	CAT, CAC
START "("	ATG
Isoleucine	ATT, ATC, ATA
Leucine	CTT, CTC, CTA, CTG, TTA, TTG
Lysine	AAA, AAG
Methionine	ATG
Phenylalanine	TTT, TTC
Proline	CCT, CCC, CCA, CCG
Serine	TCT, TCC, TCA, TCG, AGT, AGC
Threonine	ACT, ACC, ACA, ACG
Tryptophan	TGC
Tyrosine	TAT, TAC
Valine	GTT, GTC, GTA, GTG
Stop ")"	TAA, TGA, TAG

The next step in the primer would be to describe how codon triplets map to amino acids, which would be described using similar 2D or 3D notation as the nucleotides were. Once the recipient has learned this translation table, they would be able to read genetic sequences in shorthand form.

An extraterrestrial civilization that wishes to include genomic information in their transmission could follow a similar pattern to describe the chemical "alphabet" of their genetic system and how it is used to code for more complex molecules (e.g., their equivalent of amino acids and proteins).

What Could We Learn from Genomic Information?

We could learn a great deal from genomic information in an ET transmission. Even if we are unable to understand how genetic instructions translate into function, we may be able to understand the molecular alphabet of the system, how information gets passed along between generations, and, from that, how similar or different life and the pathways to life are on other worlds.

Abiogenesis (How Life Started)

One of the big questions in science is how life got started on Earth. We obviously weren't around when it happened, and so we are stuck with parsing indirect clues to figure this out. We know pieces of the puzzle, for example, that amino acids can be produced from simple gas mixtures exposed to electrical discharges and that amino acids also exist in outer space. It appears that many of the chemical constituents of life are abundant and are produced through processes that should be commonplace throughout the universe.

Yet there is a big gap in knowledge between how amino acids and other building blocks were formed and the emergence of the first cell that was capable of encoding genetic information and reproducing itself. More importantly, we don't know if the gap between amino acids and reproducing cells is a long chain of one in a million events, with exceptionally long odds of every event along that chain happening, or if it was basically inevitable once the right conditions were in place. Our knowledge about the actual conditions and chemistry at the time life got started on Earth is incomplete at best. We also don't know if life started independently on Earth or if it started somewhere else and was later transplanted here.

Other civilizations could tell us a lot about how life got started on other worlds, whether this is an exceptionally rare or common event and whether there are only a few pathways to life or multitudes. One would imagine that other civilizations will also be curious as to how life got started elsewhere and that this is the type of information that will be worth exchanging as part of interstellar discourse.

Panspermia

One of the theories of the origin of life is that it was seeded throughout the galaxy, whether by design or by accident. This theory, known as *panspermia*, posits that once life gets established, it can be transported to other sites and spreads farther over time.

This is certainly a possibility within a solar system, as asteroid impacts routinely transfer materials between planets. The Allan Hills meteorite, which we discussed in C Day, was discovered in Antarctica which had originated from Mars. For a time it was thought to contain fossilized Martian bacteria. Bacteria and other microorganisms inside of a meteorite could survive space travel and atmospheric entry. It's plausible that life from one planet can end up seeding life on neighboring worlds within a solar system and without invoking the need for a higher intelligence or spacefaring civilization to make that happen.

Interstellar panspermia is another matter. While asteroidal transfer among planets within a solar system should be commonplace, it is less likely that random collisions like this will deliver matter across interstellar distances. The energies required are much greater, and the chance that a piece of rock chipped off one planet will travel light-years and just happen to land on a planet orbiting another star will be much, much smaller (not impossible, but very unlikely). If panspermia is occurring on an interstellar scale, this will probably be directed by a spacefaring intelligence.

Genomic information from other civilizations would help us figure out whether life had spread across interstellar distances via panspermia. If this had happened, we would expect to see the following things:

- Life in the majority of worlds would have a common mechanism for recording and transmitting genetic information. If we see that most civilizations on the network are inhabited by organisms that use RNA or DNA and use the same base pairs, this will be an indicator that panspermia may have been at work.

- There should be segments of ancestral genetic code that appear on all of these worlds and are passed down to species as they evolve from a shared progenitor. This would be the big tell – a block of genetic code that appears in organisms not just on Earth but also on many other worlds.

If we were to discover that a common genetic sequence on Earth, such as the 16S ribosomal DNA sequence, had appeared in the genomes from other worlds, this would be a pretty clear sign that there was a shared genetic ancestry between them. This sequence is responsible for the synthesis of proteins from chains of amino acids and is found in all bacteria; a related form is found in higher organisms.

Convergent Chemical Evolution

Another possibility is that life on most worlds uses RNA or DNA to record genetic information, not because of panspermia, but because these happen to be the most stable and energy-efficient ways to store genetic information. For instance, we might discover that life on other worlds uses these same systems, but there would be no sign of shared genetic sequences across worlds. So while they use a common system for storing information, there would be no overlap in the encoded information itself. This would not be to say that other genetic encoding systems are impossible; it's just that in most situations, RNA and DNA win out over the long term and become the primary mechanism for transmitting genetic information on most worlds.

Diverse Genomic Systems

Another possibility is that DNA and RNA are just one of a countless number of ways to record and transmit genetic information and that their emergence on Earth was just an accident of history. In this scenario, there would be nothing special or magical about DNA. Each world would have evolved its own system for recording genetic information that is well adapted to its chemical environment.

Imagine that we find ourselves in contact with a network of civilizations and that each of them is sharing genomic information from their world. If the system used to record genomic information differs from world to world, we'd expect to see a different primer for each system that describes how information is recorded, for example, which molecules encode for different amino acids or similar building blocks.

We might discover that molecules like RNA and DNA are favored on temperate, watery worlds like Earth, but that something completely different wins out on worlds that are much hotter or colder than Earth, or that there are systems that can function in conditions that we don't think of as being hospitable for life. Saturn's moon Titan, for example, is covered with super cold hydrocarbon oceans. Life on a world like that could have evolved around a much different genetic system.

The Tree of Life and the Formation of Species

On Earth, species fork off from each other like branches on a tree. Except for bacteria, genetic information is generally not passed laterally between species, so each branch and the branches that form from it remain separate from the others once they are formed. This pattern of speciation forms a tree of life that allows us to trace a genetic history back in time.

Perhaps this is a universal pattern, or perhaps it is specific to Earth. In Earth life, genetic information is passed along from parents to offspring and does usually not make lateral jumps between unrelated individuals, except in the case of some bacteria. It's possible that on other worlds, the concept of a species is elastic. On a world where genetic information travels more freely, we would see genetic sequences that make lateral hops between individuals or species, much like bacteria on Earth can exchange genetic information via plasmids, loops of DNA that can be passed among them separate from reproduction. This could lead to a very different pattern of evolution, where the development of an advantageous trait in one species could spread to many others. Instead of a competition between species, there would be competition among genes that are widely shared among species.

Patterns of Reproduction

If the genomic samples include genomes for individuals, it may also be possible to learn how information is passed through reproduction between generations. In Earth life, asexual reproduction is common among fast-reproducing organisms like bacteria. Bacteria make so many copies of themselves that random mutations can drive evolution on short time scales. Long-lived organisms, such as animals, rely on sexual reproduction to mix genetic traits in offspring.

The sender could include genomic information for a large population of individuals, especially if they use inscribed matter as part of their system. One kilogram of inscribed matter, with an information density of 10^{22} bits/kg, would be able to encode the genomes of over 10 trillion human beings. If they are limited to transmitting information via radio or optical signals, they may be somewhat limited in how much genomic information they can send. Even then, it will be possible to transmit complete genomes for a representative sample of individuals should they choose to do so.

A Comparison of Genome Sizes

How much information will it take to send the complete genome of an animal? While we don't know about life on other worlds, we can look at the genomes of Earth organisms to get a sense of how genome size varies from single-celled organisms to humans, as shown in Table 16.2.

If an ET civilization chooses to include genomic information in their transmission or via inscribed matter, they could easily do so. Were we to include this information in our outgoing transmission or inscribed matter probes, we could send not only the genomes of many species but also the genomes of entire populations. If we wanted to preserve the genetic history of human beings on Earth, we could send the genomes of every living person using just a few grams of inscribed matter.

Table 16.2 A comparison of genome sizes

Name	Latin name	Genome (bits)	Time to send (1 megabit/sec)	Inscribed matter(10^{22} bits/ kg)
Algae	*Ostreococcus lucimarinus*	26 megabits	26 seconds	2.6 picograms
Fruit fly	*Drosophila melanogaster*	260 megabits	4 minutes	26 picograms
Horse	*Equus ferus caballus*	5.4 gigabits	1.5 hours	< 1 nanogram
Human being	*Homo sapiens*	6 gigabits	1.6 hours	< 1 nanogram

The History of Life on Other Worlds

One of the more useful aspects of genomic information is that it can be used to work out the history of how life evolved. Geneticists do this by tracking how DNA sequences change over time. As organisms reproduce, they pick up mutations – changes to their genetic code. These mutations happen at a fairly predictable rate, and so by measuring how much a genetic sequence has changed between two different organisms, it is possible to work out how many generations have elapsed between them. Some sequences are particularly useful as genetic clocks, because they appear in almost all organisms and have known mutation rates. By studying how these shared blocks of code differ between sequences, we can learn which species they were inherited from as well as their ancestral heritage.

In Vivo Experiments with Alien DNA and Why They Will Be a Bad Idea

While it's one thing to analyze genomic information on a computer, it would be another thing entirely to introduce that information into living systems here on Earth. This is such an obviously bad, awful, no-good-could-possibly-come-from-it idea that it may be inevitable that somebody somewhere will decide to try to do this.

The risk would be particularly acute if we find that alien life, like on Earth, encodes information in DNA and shares the same basic amino acids. If that were the case, someone could use a technology like CRISPR to insert genetic information transcribed from an alien transmission into Earth organisms and see what happens.

The risk probably (key word: probably) is not so much that an errant experiment will introduce an alien super-virus, but rather than it could introduce what turns out to be an invasive species. Imagine that an alien genome contains genes for a more efficient process for photosynthesis. This new process isn't radically more efficient than it is in Earth life-forms, just slightly better. The problem is that small differences in efficiency compound over time, so even a slight advantage can become overwhelming over generations. So the risk is that a seemingly innocuous experiment with algae could turn out to be a sort of kudzu from hell that is capable of wrecking entire ecosystems.

Hopefully if and when someone decides to do an experiment like this, they'll do it in a space-based robotic lab, on the far side of the Moon, with no humans ever setting foot in it, with at least one high-yield hydrogen bomb attached to a dead-hand switch just in case things go wrong.

Preservation

One of the most potent motivations for a civilization to include genomic information might be a combination of vanity and self-preservation. We don't know yet what the average lifetime of a civilization is, and our own experience to date isn't very encouraging. Even long-lived civilizations may go extinct, and as they face that eventuality, they may want to provide the universe with a record of their existence and perhaps the means to reconstitute their biosphere or parts of it somewhere else.

Consider that we could encode the genomes of over 10 billion people in a gram (1/30th of an ounce) of inscribed matter. We could build an interstellar probe and use it to deliver genomic data not just about human beings but about most of the plants and animals on Earth. We could then launch something like the New Horizons probe and put it on a trajectory that would get it to a nearby star within a few tens of thousands of years, which is nothing in terms of the lifetime of the galaxy. The challenge for us would be less in launching the space probe and more in collecting so much genomic data from people, animals, and plants, though our ability to sample this information is growing exponentially.

If we are capable of doing something like this with current-day technology, it's reasonable to assume that a more advanced civilization will be capable of doing so if they choose to. So not only could we find ourselves receiving scientific and cultural information from another civilization, but we could also find ourselves receiving information about alien life itself.

17

The Galactic Internet

We should consider that an ET transmission may be part of a larger network. The detection of one nearby civilization would mean it is likely that there are others and potentially many others. The main question at that point is whether the others are close enough to each other for multipoint communication to get started.

A *mesh network* is an energy-efficient way to transmit information over very long distances, by using sites in between endpoints to relay information, versus direct point-to-point communication. If intelligent civilizations are common enough, for example, if there's one every 100–1000 light-years, a mesh network would reduce the energy cost of communicating across galactic distances by several orders of magnitude.

In Lessons From Computing And Communications, we see that the steps needed to segment data and protect it from damage due to errors are similar to the steps needed to build something like a mesh network.

The big if is whether communicative civilizations are densely distributed enough that ongoing communication gets started among more than two sites. If that happens, there are a number of economic arguments that point toward the development of a system with the characteristics of a mesh network:

- **Energy efficiency** – mesh networks yield multiple order of magnitude reductions in the energy required to transmit information versus direct point-to-point communication, especially across galactic distances.
- **Reliability** – in a mesh network, there are many paths between any two nodes, which increases reliability. Information can be routed via

B. S. McConnell, *The Alien Communication Handbook*, Astronomers' Universe,
https://doi.org/10.1007/978-3-030-74845-6_17

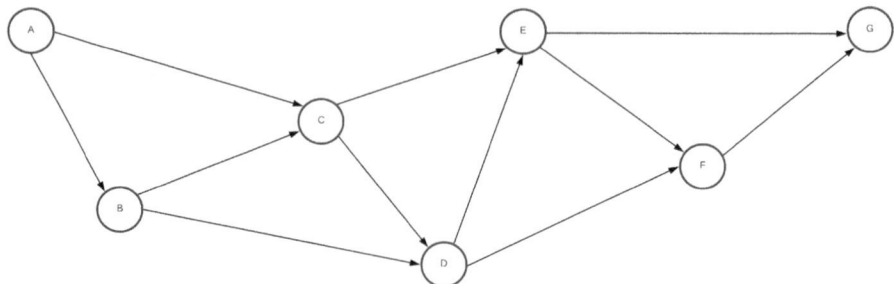

Fig. 17.1 A simple mesh network with seven nodes. Notice that there are many paths between node A on the left and node G on the right. (Image credit: Brian McConnell)

more than one path, thereby preventing an outage in one point to point link from causing the loss of connectivity or data.

- **Beam targeting** – at great enough distances, a transmitter needs to forecast its target's position far into the future and aim its beam at that future location. Nearby stars will have moved less and will have done so less unpredictably, compared to stars at galactic distances.

- **Short legs** – the amount of time required for information to travel between any two nodes is minimized, which will reduce latency in the network.

- **Selection pressure toward intelligibility** – the transmission that is easiest to parse compared to other attempts at interstellar communication is more likely to be repeated and mimicked, which is not unlike Darwinian selection and reproduction. This would favor a system that is easy to parse yet also able to communicate a wide range of media types.

- **Distribution and preservation of important technical and cultural information** –information in a large-scale mesh network like this could be archived at multiple sites and recirculated in perpetuity, including information from long-extinct ancestor civilizations.

- **Collective defense** – if there are hostile civilizations, perhaps it is better to be part of a larger network so one can gather information and warnings about potential threats.

This is speculation, of course, but should nevertheless be considered. There are several things to look for in evaluating the data in an extraterrestrial transmission for clues about the nature and scale of the communication taking place.

If the data is segmented, the metadata attached to each parcel of data should hint at the size of the network. In a point-to-point connection with no other nodes, a sender ID and receiver ID are redundant information that would just cut into information carrying capacity. If this metadata is present, the size of these address fields will hint at the size of the network.

It is helpful to have an idea about the network topology and who the other sites are. For example, a sender that is aware of other sites that may be within detection distance of us would be able to include information about where to look for additional nodes, as well as caches of inscribed matter that had been delivered to our vicinity. It may take some time to figure out how they are trying to point this out to us, but we should be looking for this sort of information.

If the transmission is rich with images, look for visual cues that the scenes shown are from different perspectives and cross-check that with whatever metadata is available. Images of mutually observable objects taken from different solar systems would have noticeable parallax differences compared to foreground objects, which would reveal the relative locations from which they were taken.

This type of network might also involve multiple information-delivery mechanisms. Electromagnetic carriers will be limited in the amount of information they can transmit in a given time period, even if the usable spectrum is fully utilized. As discussed earlier, a microwave link would probably be able to deliver at most about 10^{17} bits per year. While this is a lot of information, it is tiny compared to the amount of information that could be exchanged within a network of advanced civilizations.

This will favor the use of inscribed matter to deliver archival information and information from distant nodes in the network, while reserving the electromagnetic carriers for transmitting information that is more time sensitive.

If we were to find ourselves in contact with a network like this, contact with one civilization could mean contact with many civilizations, carrying truly profound implications.

Checklist of Patterns to Watch For

Is there a mechanism for indicating where a particular block of data originated from, the equivalent of a sender ID? If so, how large is this address space? How many unique addresses do we see in the data stream?

Is there a mechanism for indicating when a block of data was sent (an equivalent to universal coordinated time)? This might hint at how long the network has been operating or how long a particular block of data has been in circulation.

Is there a mechanism for indicating who a particular block of data is intended for, the equivalent of a receiver ID? If so, how large is this address space? How many unique addresses do we see in the data stream? Does there appear to be a receiver ID assigned to us?

Are there images or other observational data sets that appear to be taken from different sites? For example, are there images of the same nebula taken from different solar systems with noticeably different parallax angles?

Are there diagrams or other data sets that appear to indicate where other nodes on the network may be physically located or what the network topology looks like (e.g., something similar to the pulsar spin interval map used in the Pioneer and Voyage records)?

18

The Message Analysis and Comprehension Effort

While the initial stages of signal detection and analysis will mostly involve SETI scientists and subject experts, the process of understanding what another civilization has to say will involve a large and diverse group of people. It is likely to become the largest scientific undertaking in history because of the number of people who will be able to participate and because anyone with a computer and Internet access will be able to examine the data from the transmission.

As the message analysis and comprehension effort proceeds, individuals and teams will develop utilities that assist others in working with the data being received. Thanks to the Internet and distributed software development services like Github, these tools will probably be widely shared soon after they are developed. This will enable subject matter experts to focus on the specific topic they are interested in, such as interpreting pictures of landscapes, without needing to understand other aspects of the message, such as low-level formatting or encoding.

It is important to point out that at the beginning of the process, we will have no idea how the message is organized or what types of media it conveys. The goal as we develop analytic tools and utilities will be to automate steps in this process, such as converting raw data into a user-friendly format that maps well to the computing and analytic tools used for higher-level analysis. You can think of this as a processing pipeline that will enable users to automate many aspects of the analysis work as we learn more about how the communication system is organized.

B. S. McConnell, *The Alien Communication Handbook*, Astronomers' Universe, https://doi.org/10.1007/978-3-030-74845-6_18

The Interstellar Communication Relay

One of the initial steps in the message analysis and comprehension process will be to build a repository to store and distribute the raw and derived data products the receiving telescopes generate. The system will serve several different user communities, as illustrated in Fig. 18.1, each with different needs and capabilities, and will likely be organized around each use case. The primary goal of this system will be to ensure that anyone who wants to have access to the received information will be able to do so and to create a data-processing pipeline that allows people with different skill sets to work on different aspects of the analysis and comprehension effort.

Tier 0: Raw Data Access

Astronomers and signal-processing experts will want to examine the raw data from observing telescopes to look for additional carriers and modulation schemes. This work will require them to process large amounts of data. This equates to 10–100 gigabits per second for a radio telescope observing a wide swath of the microwave spectrum, more data than can be carried even by ultra-high-speed Internet connections, which is further complicated by the fact that many of these facilities are in remote locations. It is likely the observing facilities will grant on-premise network access to guest researchers, who will bring in their own equipment racks to run their data-processing pipelines. These in turn will produce derived data products that may be shared with downstream users.

The number of people who will require this level of access and who will have the skills needed to do useful work will be small, so this need can be

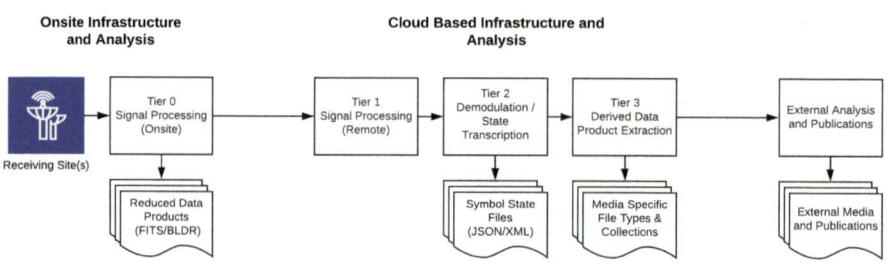

Fig. 18.1 The data-processing pipeline for the Interstellar Communication Relay. (Image credit: Brian McConnell)

served by on-premise equipment combined with "sneakernet" data transport to locations with better connectivity for offsite storage and data access.

Tier 1: Derived Signal Data

This raw signal data will be reduced to discard data that is not needed (e.g., frequencies where no signal is present). The information will still require significant storage and bandwidth but will likely be on the order of tens or hundreds of megabits per second, which is possible to transmit over a fast Internet connection in real time and then archive in a cloud computing service such as those offered by Amazon or Google. The users for this type of data product will be signal-processing experts, people experienced in building communication systems, etc. The ability to work with data offsite will enable larger numbers of people and institutions to participate in this effort.

Tier 2: Demodulated Data Streams

If the signal is modulated, the data extracted via the demodulation process will be made available in real time or near real time. This data product will require orders of magnitude less bandwidth and storage and can be made accessible via cloud computing services.

The demodulation service would transcribe data into an easily parsed format so downstream users can process it however they want. No attempt is made to understand the contents of the data stream at this stage in the pipeline.

The users for this data product will be people who want to test different assumptions about how the data is organized or structured, whether the data is segmented, what types of error correction are in use, and so on. This will be accessible to a large community of users – basically, anyone who is familiar with computer programming will be able to participate in this effort. A system like Github, a popular system for hosting software development projects, will work well as a repository for this data.

Tier 3: Structured Data Extraction

This step in the pipeline will process the demodulated data flowing from tier 2 streams and, as we learn what data structures and types it contains, will extract higher-level data types, such as images, and write these out to something that looks like a computer file system, as that is what most users will be familiar with. A system like Github will likewise be useful as a repository for these data products.

This step in the pipeline will transform these media types into formats that are familiar to computer users, such as PNG or JPEG image files, WAV audio files, and JSON for structured data, as well as scientific file formats like FITS. This will enable subject matter experts who may not be expert programmers to work with the data products.

The community of users for this data product will expand to a wide field of disciplines, especially if observational data such as images and sounds are part of the transmission. Biologists, archeologists – every field you can think of – will be interested to see what is there and what is being communicated.

Separation of Concerns

Before we discuss the specific steps in the analysis and comprehension effort, it's good to talk about the concept of *separation of concerns*. This is an often-used concept in computer systems and software design. The basic idea is to separate systems or processes and design well-defined interfaces between them, so that someone working on one component of the system only needs to know how the interface to another system works and does not need to know about the other systems' inner workings. This frees people to focus on their area of expertise and not get bogged down in the details of other peoples' work. The tools and systems built to support the analysis and comprehension effort should be built around this principle.

Let's imagine that the transmission contains a large number of images, many of which are photos of alien structures or settlements. People from many different fields of study will be interested in these images – biologists, archeologists, and architects, to name a few. Most of these people, while they may be comfortable using photo-processing software, will not be software engineers or image-processing experts. The analysis tools should enable people to work with data products that are well matched for their needs and skills.

Initial Analysis and Visualization

Before we even begin trying to understand the contents of the data received, we will want to understand how the data is organized, if it appears to be segmented and so on. It will be helpful to have tools that allow people to visualize and inspect the raw data for patterns and structure. Examples of the types of utilities we would like to have for this stage of analysis include:

A visualization tool that displays the raw data as a bitmap image and allows the viewer to adjust parameters such as pixel depth, row length, number of color channels, and so on. This will allow for visual inspection of the raw data. If repetitive patterns such as padding are common, these should be readily visible.

A tool that maps the Shannon entropy of the data set, to reveal regions with high or low information density. This will be useful in identifying regions of the data set that are highly structured or repetitive relative to other sections. See Chap. 8 Entropy for examples of this type of analysis.

A crowdsourcing tool that enables a large number of Internet users to participate in looking for patterns in the data set.

People are adept at recognizing patterns and can be recruited to help in parsing the structure of the data set. A crowdsourcing tool would display random segments of the data set and ask the user for each view if they see a repetitive or structured pattern. This will enable researchers to identify interesting regions of the data set, which can then be subject to more thorough scrutiny.

Because a detection event will generate a high level of public interest, it will be possible to attract millions of volunteers to a project like this. In fact, the SETI@Home distributed computing project attracted several million volunteer users during its 25 years of operation. With this level of participation, a system like this will be able to direct tens of thousands of person-years to the task of looking for patterns and structures in the data set.

It should also be possible to build AIs that analyze incoming data to look for specific patterns, such as those we might expect in bitmaps. Let's imagine that we have received a gigabit of data from the transmission and want to see if there are images somewhere in this data set. An AI algorithm that has been trained to recognize images could be used to check a large number of guesses at decoding images and, for each decoding attempt, return a "may be an image" or "is not an image" result. Incorrect guesses at encoding parameters, such as the number of bits per pixel, will generally result in an image that looks like static. Special-purpose AIs such as this should enable us to grind through a large number of permutations for decoding images, audio, and other media in a short time.

Low-Level Decoding and Collection Extraction

One set of tools will be focused on converting raw transcribed data into an intermediate format that is easy for people to work with. Suppose the message is composed of many small blocks of data, each of which is labeled with meta-data to identify which larger objects and collections it belongs to. One of the first utilities that will be developed is a program or library that processes raw data and transforms it into something that looks like a collection of files in a computer file system or directory tree. Once built, most analysts will not need to be concerned with the low-level encoding system and can instead work with the output from this step in the processing pipeline.

One of the important goals at this stage will be to write software that can translate the raw data into a format that maps well to the computing systems and tools we use, as this will enable a wide range of people to work with the data effectively. This set of utilities would also deal with tasks such as error detection and correction. People who are experienced with operating systems design and software architecture are likely to play a significant role in this stage of the analysis effort.

While it is impossible to predict the specifics of how such a message might be organized, the principles of communication network design will favor certain features, such as message segmentation and forward error correction.

Media Extraction and Conversion

Let's consider an example where the sender is transmitting a large collection of images. Some of them are panchromatic (grayscale) images. Some are multi-channel color images. Some are stereoscopic sets of images used to build 3D models. In the previous step, a program would transform the raw data into what looks like a collection of folders and files.

Once we have learned how to decode images reliably and discern between different image types, it will be helpful to have a utility that automatically converts these into image formats that are widely used on computing devices and the web. A lossless format like PNG is a good candidate as an output format for images because it is universally supported across devices and web browsers. A format like FITS will work well for scientists. Similar utilities could also do conversions to translate images into human-centric color models for casual use, while others would be used to generate files for scientific analysis. The goal at this stage would not be to understand the contents of images, but just to reliably convert the native image format into formats that are easy for less technical users to work with.

If audio samples are part of the message, they would likely be stored as linear (time domain) sets or as frequency domain (spectrograph or waterfall plot) files. Ideally what we would like to build in this stage of analysis is a set of utilities that can recognize the type of encoding and automatically transform this into a widely used audio format such as WAV, which can be played on virtually any device. Once built, it would be possible to extract audio into a large number of files that other analysts could go to work at interpreting, again without needing to understand the low-level encoding details.

N-Dimensional Models

Multi-dimensional structured data, such as arrays, point clouds, etc., are also something we will want to be able to recognize and process automatically. A 3D point cloud might be described as a list of three-dimensional coordinates, in the form:

$$\left(\left(\left(x_1\right)\left(y_1\right)\left(z_1\right)\right)\left(\left(x_2\right)\left(y_2\right)\left(z_2\right)\right)\left(\left(x_3\right)\left(y_3\right)\left(z_3\right)\right)...\left(\left(x_n\right)\left(y_n\right)\left(z_n\right)\right)\right)$$

Analysts will study the data to look for markers used to represent structure, for example, the equivalent of parentheses. Once we learn to recognize these, it will be straightforward to write programs that parse this data and transform it into formats used in day-to-day programming and 3D-rendering applications (see Chap. 11 Three-Dimensional Images and Models).

Algorithmic Communication Systems

If the transmission includes algorithmic components, we will want to build virtual machines, or simulated computers, that can run these programs. To do so, we will need to understand the basic instruction set that the programs reduce to.

As discussed previously, there are a number of ways to build an algorithmic communication system, which include (1) a logic gate network that describes the computing substrate in explicit detail and (2) a symbolic system that maps to the instruction set for a virtual machine. We won't know the specifics of which approach an ETI might take, and they may take several since some approaches work better for certain cases than others. Computer scientists and hobbyists will be able to make important contributions to understanding this content and will be able to test many hypotheses in parallel. This is also a case where individuals and small teams may make important contributions to the analysis and comprehension effort.

Once the above is understood, it should be possible to build programs that can run these instructions in a simulated environment and to probe their behavior and outputs.

If the algorithms are written in an interpreted programming language, we will need to build an interpreter to run them. Once we understand the basic instruction set, this should be straightforward to do and can be done by an individual contributor or small team.

If the sender instead describes the underlying computing substrate, we will need to build a program that simulates arbitrarily complex logic circuits or transforms the statements in the ET data stream into a format that can be read by popular electronic simulation programs such as MATLAB.

Semantic Networks

A semantic network will enable the sender to build a knowledge graph that contains a large number of concepts, identified by unique numeric codes, and that can be described in terms of their relationships to other concepts. As discussed earlier, this type of system can be used to create an artificial or constructed language and will enable the communication of a wide range of ideas.

A semantic network can be built using a series of expressions that will take a form such as the following example:

```
((symbol_number)(operand or expression)(symbol_number))
((symbol_number)(operand or expression)(symbol_number))
```

   ...

The numbers assigned to individual symbols or operands may be arbitrary, so we should not expect them to convey any meaning. The first step will be to recognize the general patterns used in these statements.

The next step will be to analyze the entire data set to rank symbols by the frequency they occur and set about deciphering what the most common symbols refer to. For example, a symbol that means "is a member of set {N}" is likely to occur quite often in a semantic network. In general, the strategy should be to decipher the operands that explain how other symbols relate to each other and then proceed from there to less common symbols.

Once the basic operands are known, it will be possible to automatically build the knowledge graph for the entire network. While the meaning of the symbols in the network will not be known yet, it will be possible to see how symbols are related to others in the network. It will also be possible to build a database of every symbol encountered, its mappings, and, if deciphered, its meaning in human languages.

Deciphering the meaning of individual symbols may require a variety of techniques, depending on what the sender is trying to communicate. A straightforward example would be to associate symbols with images contained elsewhere in the transmission, for example, to build up vocabulary related to real-world objects or processes. As we decipher symbols, the translation table that maps numeric identifiers to human language forms would be updated. This in turn will aid in deciphering other symbols that are described in terms of symbols that have already been deciphered.

A semantic network can also form the basis for an expert system that can answer queries about symbols in the network. This may be particularly useful because even if a symbol's meaning is unknown, it may be indirectly linked to other symbols that have been successfully decoded. The expert system would respond to queries by returning a map of the relationships to the symbol being queried.

Let's consider an example of an unknown symbol that is mapped to others in a small, partially deciphered network. The network describes relationships between symbols related to several types of organisms and consists of the following statements:

```
((72934) (211) (120038))
 ((718239) (211) (72934))
 ((718239) (210) (63242))
 ((617298) (210) (63242))
 ((617298) (211) (72934))
 ((43826) (211) (72934))
 ((43826) (245) (4500))
 ((120080) (245) (4500))
 ((120080) (211) (120038))
```

Some of these symbols have already been deciphered as follows:

210 means "has"
211 means "is a member of a set"

245 means "lives in"
4500 means "water"

So we can substitute these operands to partially translate the network expressions as:

72934 is a member of 120038
718239 is a member of 72934
718239 has 63242
617298 has 63242
617298 is a member of 72934
43826 is a member of 72934
43826 lives in water
120080 lives in water
120080 is a member of 120038

We can graph this network as shown in Fig. 18.2 and Table 18.1.

Even with a partial understanding of what symbols mean, we can get an idea of how they are related to others, for example, to understand what categories they belong to, what attributes are associated with them, and so on. As previously discussed, the numeric addressing system is particularly useful because the sender can ensure that each node in the network has one and only

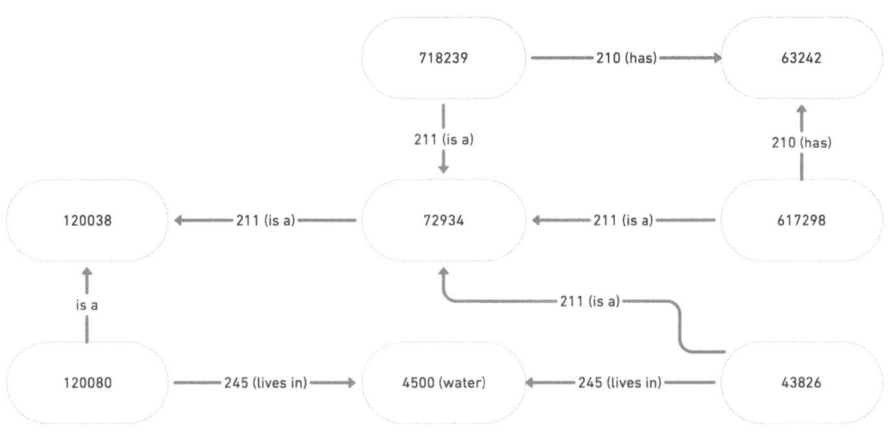

Fig. 18.2 A diagram of a partially understood semantic network. Notice that even though only a few of the concepts are mapped to human language terms, we can still understand how the concepts are connected to each other. (Image credit: Brian McConnell)

Table 18.1   A sample list of semantic network addresses and their meanings

Symbol ID	Meaning
210	Has
211	Is/is a member of
245	Lives in
718239	Dog
617298	Bear
120038	Animal
120080	Fish
4500	Water
72934	Mammal
43826	Whale
63242	Fur

one meaning. This is different from human language, where a word may have many different meanings depending on its context. A semantic network can prevent this type of confusion and also define a very large vocabulary of concepts. This type of system also enables the user to build a map of unknown symbols.

Genomic Information

Genetic information, such as a DNA sequence, can be represented in a digital medium. This is another type of data we should be watching for. A pattern we would be likely to see is a small group of symbols that are grouped in small n-grams (pairs, triplets, 4-grams, etc.). RNA and DNA, for example, have four base pairs that are grouped into triplets, which allows for 64 possible combinations in each genetic word.

As we discussed in Genomic Information, there are a range of possibilities we might encounter. It may turn out that DNA is a common encoding system, either due to panspermia or because it generally wins out as the most efficient way to store genetic information. On the other hand, we may find that each world evolves its own system for storing genetic information and that there is nothing special about DNA. Either one would be an important finding, as it would tell us a lot about how life has developed elsewhere.

This information will probably appear to be pretty random when its Shannon entropy is measured. To understand what it represents, we will probably need a primer that maps molecular structures to numeric symbols that are used as shorthand (see Jon Lomberg's primer on DNA from the Voyager Golden Record for an example, which we discussed in Chap. 16 on Genomic Information).

Once we have learned to make this association, we would then be able to read genome sequences from the data stream. Geneticists and life science researchers would play an important role in analyzing this information and in adapting existing software to work with these data sets. For example, if we were to receive genomes from many related organisms, we would be able to compare them and look for differences that reveal details about how information is shared, including heredity, and evolution.

Other Data Types and Sensory Modalities (Unknown Unknowns)

We are limited by our senses and experiences, and that will inevitably bias us to look for certain data types while ignoring others. While one can argue that certain data types, such as images, are likely to be universal due to their importance in astronomy, we should be careful to explore all possible modes of representation.

Let's imagine a technological species whose primary mode of vision is via echolocation. How might they record and share their "pictures" with others? Would they record echolocation photos as some sort of processed image or data product? Or would they record the outgoing sounds and return echoes in a raw format and assume that the listener would be able to form a mental image from those sounds? It's hard to say what sort of data products a species like this would prefer and what sorts of tradeoffs they might make when building a digital communication system around them.

This is why it is important that the analysis and comprehension effort be open to anyone who wants to participate, as it is possible that some of the most important contributions will come from diverse people who are not affiliated with SETI and who may approach these problems from very different perspectives.

19

What Could We Learn from Another Civilization?

A digital communication system will be capable of conveying many different types of information that can be combined in many ways. What another civilization will be able to say with a system like this will only be limited by their creativity.

Because a digital communication system can combine many different types of information, the sender will not be limited to a single medium or mode of communication to get their point across. Instead, they will be able to use the media that are best fit for the goals they are trying to accomplish. They will also be able to combine media that require different levels of skill and intellectual capacity to comprehend and will be able to design messages that can be at least partially understood by a wide audience, hopefully including newcomers like us.

As we discussed, qualia, or internal experiences, may be very difficult to communicate to someone who lacks shared physiology or experience. Communication built around observables or representations of objects or processes should be much more approachable.

Virtual Exploration and Experiential Communication

Interstellar communication, especially if it is occurring in the context of a long-lived network of civilizations, could enable civilizations to explore interstellar destinations without leaving their home systems, and do so at much less expense

© The Author(s), under exclusive license to Springer Nature Switzerland AG 2021
B. S. McConnell, *The Alien Communication Handbook*, Astronomers' Universe,
https://doi.org/10.1007/978-3-030-74845-6_19

and risk. By combining images, audio, three-dimensional models, and other media, a sender could enable receivers to experience environs from their world.

Imagine that a civilization wants to share a first-person view of an important city or settlement. This could be done by sharing observables (pictures, audio samples, and 3D models), all of which can be represented numerically. It might take the receiver a while to figure out the particulars of the different media types and how they are grouped into collections, but there is no need for the receiver to understand the internal experience of the author who created these collections.

A simple way to do this will be to combine panoramic (wraparound) images, such as the Mars panorama shown in Fig. 19.1, or video sequences with audio samples. This is all that is needed to provide someone with an interactive, first-person view of a scene. The same basic design pattern can be used to present a first-person view of any scene, be it an astronomical image or an up-close view of a cityscape.

What places might we be able to explore like this? Any scene you can imagine really. This is the powerful thing about photographs, that they can be used to represent scenes ranging from microscopic to cosmological scales.

Poor Man's Virtual Reality

Let's imagine that we are composing an interstellar message and want to provide immersive, first-person views of scenes ranging from cityscapes to natural environments. For each scene, we will need the following:

Several panoramic photos, each taken from slightly different positions, so the receiver can create stereoscopic images or recover depth information from the photos

A collection of audio samples recorded at high sample rates to capture audio at a wide range of frequencies (and beyond the range of human hearing)

Because a panoramic image is intended to capture the entire environment around the viewer, it will probably be at a higher resolution than an image that is zoomed to fit a specific object. For the purposes of estimation, let's assume that each image is 100 million pixels and that we use several images taken from slightly offset positions to provide parallax or depth information. We also use several color channels, so the receiver can reproduce a color scene in a way that suits them. In rough numbers, this works out to about 10 gigabits of data. This can be reduced considerably using techniques we've discussed earlier in the book, such as color downsampling, but let's use conservative assumptions for now. Audio will require much less data to represent. Let's assume that each image set is paired with ten minutes of audio samples at 100 kilohertz sample rate. This will work out to about 1 gigabit of data or about 10% of the information footprint of the imagery.

Even if the sender is limited to a relatively slow link at, say, 1 megabit per second, it will be possible to transmit several thousand scenes like this per year, and that is assuming no compression techniques are used.

Fig. 19.1 A panoramic image of the Glen Torridon region of Mount Sharp on Mars. This type of image allows the user to enjoy a panoramic view of a scene (the image appears distorted to fit within this page). (Image credit: Curiosity rover/NASA/JPL-Caltech/MSSS)

Art and Culture

Much of this book has focused on the technical aspects of interstellar communication, such as the mechanics of encoding images and sound in a digital medium. These encoding techniques will work for any sound or image, not just literal representations of scenes. In other words, they can be used to transmit what we call art. There are many definitions for art, but a common theme among them is that art is a representation of someone's subjective interpretation of the world, such as a drawing.

Art is also a technology, and an especially important one for recording the early history of a civilization. Modern photography and audio recording are recent inventions in human history. We relied on writing, drawing, and other forms of art to record our history and culture prior to the 1800s. Indeed, cave paintings dating back tens of thousands of years record the history of early human societies.

Perhaps our affinity for art is a peculiar human trait. Or perhaps the ability to create abstract representations is a vital skill upon which technological civilizations are built. From ET contact, we can learn whether art is common and, if it is, perhaps use it as a lens to understand how others interpret their worlds and history.

The History of Life on Other Worlds

There are many ways a sender can describe life and the history of life in their world. If we were to send a transmission to another civilization, it is likely that one of the things we would include is an encyclopedia of life on Earth.

This could consist of rich descriptions that include images and audio samples, along with metadata or semantic information that provides more abstract information. Genomic information could be used to describe the genetic ancestry and evolution of organisms. The sender also need not limit themselves to a single representation, but can use all of these information types and more to describe life from their world.

The receiver would not need to understand the alien's language to view these collections. It might take them time to understand some data types, such as genomic information or metadata that describe images, but the basic representations should be accessible even in the absence of a dialogue.

Astronomical Surveys

The extreme distances between even nearby stars limit our ability to survey worlds outside of our solar system. Interstellar communication, especially if it occurs as part of a long-lived network of civilizations, could enable us to see the astronomical observations that other civilizations have collected of the worlds close to them – ones we may never be able to visit or image in detail. In turn, we have access to worlds within our solar system and can collect information that would be difficult or unreasonably expensive for distant civilizations to obtain.

Trade is a theme we have revisited several times in this book. If there is a galactic network of civilizations, the information sharing could go far beyond sharing notes with our neighbors. We are tens of thousands of light-years from the center of the Milky Way, so we can't easily take direct observations of this environment. If information from civilizations closer in to the center of the galaxy is being shared and relayed outward, we could be able to see observations of galactic features, such as the massive black hole at its center, as shown in Fig. 19.2, in much greater detail than anything we could see from here, and also see deeper back in time.

Deep Time Views of Earth and Human Civilization

It is possible that an advanced civilization will have been able to take high-resolution images of Earth, comparable to the satellite images we routinely take today, and that they have been doing so for centuries, millennia, or perhaps much longer than that.

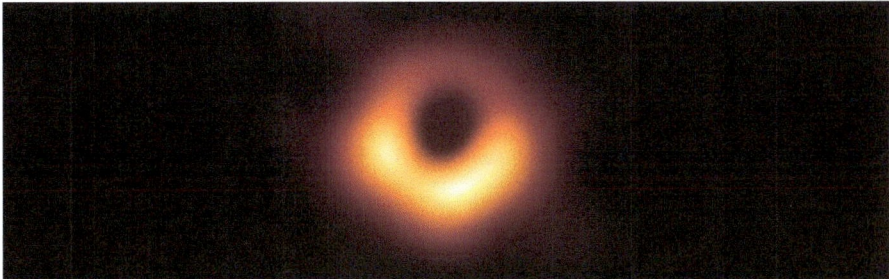

Fig. 19.2 A reconstructed image of the accretion disk surrounding the supermassive black hole at the center of the Milky Way galaxy. If civilizations closer into the galaxy's core are sharing observations, we would be able to see other parts of our galaxy in much greater detail. (Image credit: Event Horizon Telescope Collaboration)

Consider that for a moment. Contact with an extraterrestrial civilization could turn out to be a lot like the Apollo missions. We went to the Moon to learn about the lunar environment, but probably the most important cultural contribution from Apollo was the picture of the earthrise taken by the Apollo 8 crew, see Fig. 19.3.

Should we find ourselves in contact with another civilization, and especially if it turns out to be part of a larger network, it is likely we will be dealing with something more vast and more ancient than most of us can comprehend.

We are just now developing the ability to take crude pictures of planets orbiting distant stars. An ancient, astronomically literate civilization will likely have developed sophisticated remote imaging capabilities, just as are now beginning to plan robotic missions to the nearest stars that people alive today may live to see. A constellation of space telescopes orbiting at the equivalent of Mars' orbit around an exoplanet's Sun would be able to resolve small surface features on Earth and would have been able to see evidence of human activity, our cities, megalithic structures, and agriculture, possibly back to antiquity.

This is beyond our current ability, but not by much. We have already developed the analytical tools to resolve continents and oceans on distant worlds using the next generation of space telescopes now under construction. It's not unreasonable to expect that a civilization with centuries or millennia of experience with astronomy will be capable of even higher-resolution remote imaging Direct Imaging (2018).

Fig. 19.3 Earthrise, as photographed by the Apollo 8 crew, Dec 24th 1968. (Image credit: NASA)

This raises an intriguing possibility that an advanced civilization will have been aware of our existence for thousands of years, long before we developed the technology to send and detect radio or laser signals. Just as we are systematically searching for planets outside our solar system and have already cataloged thousands of them, it is not hard to imagine an ET civilization conducting similar surveys of the worlds they can see, except at higher resolution than what we are capable of today. We are already developing plans for gravitational lens telescopes that will be able to image exoplanets in detail, as shown in Fig. 19.4. A more advanced civilization may have remote imaging capabilities well beyond what we are developing today.

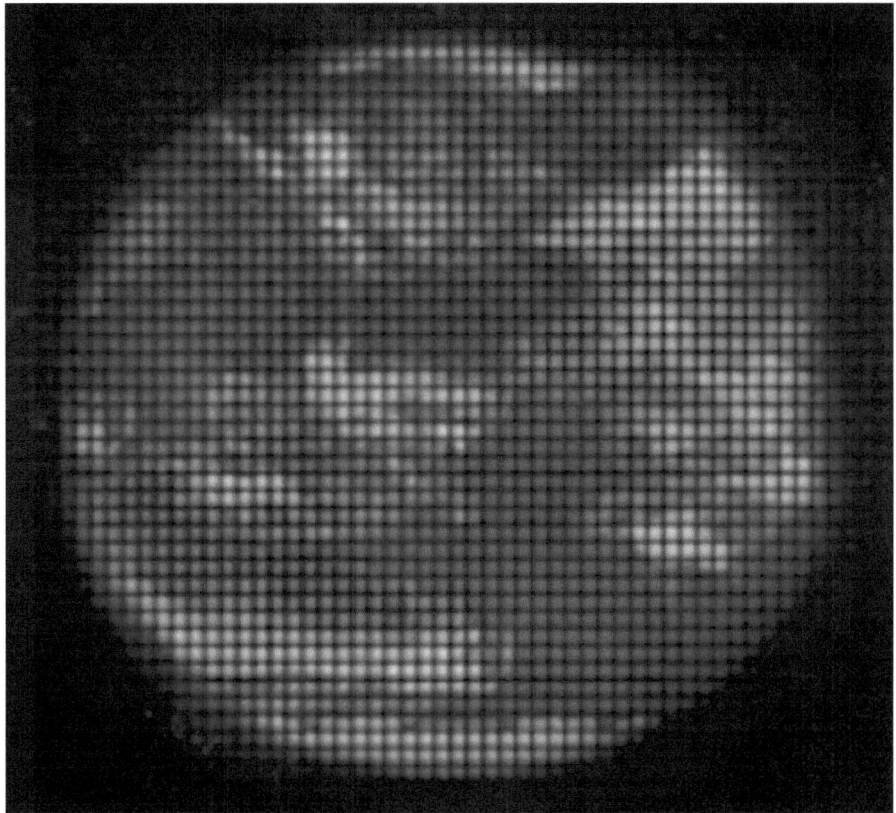

Fig. 19.4 A simulated image of an exoplanet imaged by a telescope at the Sun's gravitational lens focal line. (Image credit: NASA/JPL)

Early technosignatures of human civilization, such as large-scale agriculture and urban settlements, could have been visible to such surveys thousands of years ago, long before we began emitting electromagnetic signals. Without even intending to, we may have already revealed our existence to these astronomically literate civilizations.

If that's the case, it is possible that one of the things we will find in their transmission are images of Earth, similar to our satellite images, except dating back to antiquity. Think about how interesting that would be, to be able to peer back in deep history to see how our civilization had developed over eons. Imagine seeing an image of the Pyramids of Giza, not as they exist today, see Fig. 19.5, but as they were when they had been newly built 4500 years ago.

Fig. 19.5 An image of the Pyramids of Giza, photographed from space in 2012. (Image courtesy of NASA (Photo ISS032-E-9123 - Courtesy of the International Space Station program and JSC Earth Science and Remote Sensing Unit. https://eol.jsc.nasa.gov/SearchPhotos/photo.pl?mission=ISS032&roll=E&frame=9123))

Reference

"Direct Imaging: Svetlana Berdyugina/Jeff Kuhn", NASA Technosignatures Workshop, Sept 26–28, 2018, https://www.hou.usra.edu/meetings/technosignatures2018/presentation/?video=berdyugina.mp4.

Recommended Reading

Exercises and Sample Problems

Readers can find a set of exercises and solutions at https://github.com/alien-communicationhandbook/exercises. These exercises cover a range of topics related to SETI and message comprehension.

Selected Books, Essays, and Other Materials

After Contact: The Human Response to Extraterrestrial Life. Albert A. Harrison (1997) Plenum Books

Artificial Intelligence: A Modern Approach. Stuart Russel, Peter Norvig (2020) Pearson

Astrobiology, Discovery and Societal Impact. Steven J. Dick. (2018) Cambridge University Press

The Biological Universe. Steven J. Dick. (1996) Cambridge University Press

The Cave Of Forgotten Dreams (Motion Picture), Werner Herzog (2010), IFC Films

Chasing Doctor Doolittle, Con Slobodchikoff (2012), St. Martin's Press

Communication with Extraterrestrial Intelligence, edited by Douglas Vakoch (2011), SUNY Press

Contact: A Novel, Carl Sagan (1985), Gallery Books

Encountering Life In The Universe: Ethical Foundations and Social Implications of Astrobiology. Chris Impey, Anna Spitz and William Stoeger (2013), University of Arizona Press

Extraterrestrial, Avi Loeb (2021), Houghton Mifflin Harcourt

The Encyclopedia of Region and Society, William H. Swatos ed. (1998), Altamira Press

Extraterrestrial Languages, Daniel Oberhaus (2019), MIT Press

The Great Filter (Essay), Robin Hansen (1998), https://mason.gmu.edu/~rhanson/greatfilter.html

The Great Silence (Essay), David Brin (1983), Quarterly Journal of the Royal Astronomical Society, https://www.researchgate.net/publication/234496344_The_'Great_Silence'_The_Controversy_Concerning_Extraterrestrial_Intelligent_Life

The Great Silence: Science and Philosophy of Fermi's Paradox, Milan M. Cirkovic (2018), Oxford University Press

The Impact of Discovering Life Beyond Earth. Steven J. Dick. (2015) Cambridge University Press

Making Contact: Jill Tarter and the Search for Extraterrestrial Intelligence. Sarah Scoles. (2017), Pegasus Books

Maps of Time: An Introduction To Big History, David Christian (2004), University of California Press

Principles of Animal Communication, Jack W. Bradbury, Sandra L. Vehrencamp (1998), Sinauer Associates

Rare Earth: Why Complex Life Is Uncommon In The Universe, Peter D. Ward and Donald Brownlee (2000), Copernicus/Springer Verlag

The Society of Mind, Marvin Minsky (1986), Touchstone/Simon and Schuster

Space, Time and Aliens, Steven J. Dick (2020), Springer

When SETI Succeeds: The Impact Of High-Information Contact, edited by Allen Tough (2000). Foundation for the Future

Research Tools

In addition to the books and essays listed above, Jason Wright and his colleagues at the University of Pennsylvania have built a web search tool that tracks SETI-related papers that were published in scientific journals and pre-publication systems such as arXiv. The online bibliography can be found at https://ui.adsabs.harvard.edu/search/p_=0&q=bibgroup%3ASETI

Reference

Reyes, A., "Towards a Comprehensive Bibliography for SETI", **Journal of the British Interplanetary Society**, vol. 72, pp. 186–189, 2019.

Index

© The Author(s), under exclusive license to Springer Nature Switzerland AG 2021
B. S. McConnell, *The Alien Communication Handbook*, Astronomers' Universe,
https://doi.org/10.1007/978-3-030-74845-6